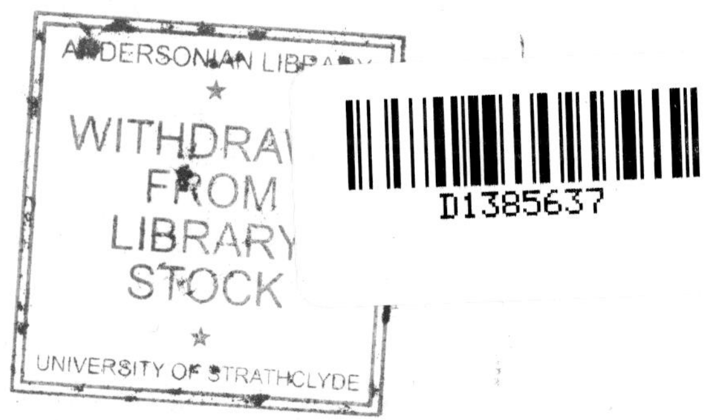
ANDERSONIAN LIBRARY
WITHDRAWN
FROM
LIBRARY
STOCK
UNIVERSITY OF STRATHCLYDE
D1385637

FEBS Federation of European
Biochemical Societies
12th Meeting Dresden 1978

*General Editor: S. Rapoport, Berlin*

VOLUME 53
Symposium S6

# PROCESSING AND TURNOVER OF PROTEINS AND ORGANELLES IN THE CELL

Proceedings of the 12th FEBS Meeting

*General Editor: S. Rapoport,* **Berlin**

Volume 51 GENE FUNCTION

Volume 52 PROTEIN: STRUCTURE, FUNCTION AND INDUSTRIAL APPLICATIONS

Volume 53 PROCESSING AND TURNOVER OF PROTEINS AND ORGANELLES IN THE CELL

Volume 54 CYCLIC NUCLEOTIDES AND PROTEIN PHOSPHORYLATION IN CELL REGULATION

Volume 55 REGULATION OF SECONDARY PRODUCT AND PLANT HORMONE METABOLISM

Volume 56 MOLECULAR DISEASES

FEBS Federation of European Biochemical Societies
12th Meeting Dresden 1978

VOLUME 53
Symposium S6

# PROCESSING AND TURNOVER OF PROTEINS AND ORGANELLES IN THE CELL

Editors
S. RAPOPORT, Berlin
T. SCHEWE, Berlin

PERGAMON PRESS
OXFORD · NEW YORK · TORONTO · SYDNEY · PARIS · FRANKFURT

| | |
|---|---|
| U.K. | Pergamon Press Ltd., Headington Hill Hall, Oxford OX3 0BW, England |
| U.S.A. | Pergamon Press Inc., Maxwell House, Fairview Park, Elmsford, New York 10523, U.S.A. |
| CANADA | Pergamon of Canada Ltd., 75 The East Mall, Toronto, Ontario, Canada |
| AUSTRALIA | Pergamon Press (Aust.) Pty. Ltd., P.O. Box 544, Potts Point, N.S.W. 2011, Australia |
| FRANCE | Pergamon Press SARL, 24 rue des Ecoles, 75240 Paris, Cedex 05, France |
| FEDERAL REPUBLIC OF GERMANY | Pergamon Press GmbH, 6242 Kronberg-Taunus, Pferdstrasse 1, Federal Republic of Germany |

---



First edition 1979

**British Library Cataloguing in Publication Data**

Federation of European Biochemical Societies.
*Meeting, 12th, Dresden, 1978*
Processing and turnover of proteins and
organelles in the cell. - (Publications; vol. 53).
1. Proteins - Congresses 2. Cytochemistry -
Congresses 3. Cell organelles - Congresses
I. Title II. Rapoport, S III. Schewe, T
574.8'761 QP551 78-41025

ISBN 0 08 023177 2
ISBN 0-08-023165-9 Set of 6 vols.

*In order to make this volume available as economically and as rapidly as possible the authors' typescripts have been reproduced in their original forms. This method unfortunately has its typographical limitations but it is hoped that they in no way distract the reader.*

*Printed and bound at William Clowes & Sons Limited Beccles and London*

# CONTENTS

Contents

## PREFACE

In the face of a flood of molecular biology research the last years have witnessed a greatly increased interest in the problems of post-translational processes and in the dynamics of cell components. It has become evident that there exists a variety of scheduled events which take place after or even concomitantly with the synthesis of peptide chains. These processes are of great regulatory importance for the cell. They include modification and processing of the primary translation product in order to make it functional or to transport it to its place of action, as well as the breakdown of cellular proteins which appears to be selective, at least in many cases. These questions are closely linked to the problems of synthesis, assembly and breakdown of the various organelles of the cell for which new powerful methods have been developed. The three sessions of our symposium which at first sight might appear to be heterogenous in subject matter are devoted to these interrelated problems.

The lectures range from the processing of virus-coded proteins by Huez to fundamental methodological aspects of the assessment of protein turnover in a complex intact mammalian organism by Garlick. There are many cross connections among the various lectures, such as between the subject matter of the first session and that of the report by Schatz, or between that of the second session and the accounts of de Duve, Grisolia and Luzikov. The editors are grateful to the contributors and hope that the present volume will provide stimulation and cross fertilization to biological and medical students in various fields.

July 1978

S.M. Rapoport
T. Schewe

SESSION I

# Post-translational Processing

# POST-TRANSLATIONAL PROCESSING OF ONCORNAVIRUS PROTEINS

G. Huez[1], J. Ghysdael[1,2], M. Travniček[3],
A. Burny[1,2], Y. Cleuter[1], R. Kettmann[2],
G. Marbaix[1], D. Portetelle[2]
Department of Molecular Biology, Free University of Brussels, 1640-Rhode St-Genèse, Belgium [1];
Faculty of Agronomy, 5800-Gembloux, Belgium [2];
Academy of Sciences, 16000 Prague 6, Czechoslovakia [3]

## I. Expression of the genetic information of oncornaviruses

The virion structure of oncornaviruses includes an inner protein core surrounded by a lipid envelope which contains at least two major glycoproteins. The viral RNA is located in the inner protein core which also contains the RNA dependent-DNA polymerase (reverse transcriptase) (1). The major glycoproteins also called "envelope" proteins have a molecular weight of $8,5 \times 10^5$ (gp 85) and $3.7 \times 10^5$ (gp 37) in the avian viruses (2). In the murine or bovine viruses, the envelope proteins have somewhat smaller molecular weights ($7 \times 10^5$ and $1.5 \times 10^5$, $6 \times 10^5$ and $3 \times 10^5$ respectively) (3, 4). These proteins determine most of the type specific antigenicity and are involved in the interaction of the virus with cellular membrane receptors.
The inner core contains four major proteins which are very similar among the various viruses from the same species (or group). These molecules carry the so-called "group specific"gs antigens. In the avian viruses two of these proteins, p15 (molecular weight 15 000) and p10 (10 000 daltons) are located near the surface of the virion particle, p12 (12 000 daltons) is closely associated with the viral RNA and p27 (27 000 daltons) is the major constituent of the inner core (5). In other viruses the molecular weights of these gs proteins are slightly different but the general structure of the virus is the same.
Each virion contains approximately ten copies of reverse transcriptase, 400-1000 copies of each glycoprotein and 2000-3000 copies of each of the gs antigens. The genetic information for all these proteins is encoded in the viral genome. This viral genome is composed of a single stranded RNA which sediments at 60-70S in sucrose gradients and has an apparent molecular weight of approximately $10^7$. This RNA is in fact segmented and can be easily converted by heating to two apparently equivalent molecules of 30-40S (depending on the virus) (6). Each 30-40S RNA subunit is polyadenylated and capped (7),

resembling in that most eukaryotic messenger RNA.
Genetic studies have shown that the avian sarcoma virus 30-40S RNA molecule (which is the most well known) contains four genes : "gag" coding for the gs antigens (total molecular weight 76 000); "pol" coding for reverse transcriptase (molecular weight 100 000), "env" coding for the envelope proteins (molecular weight 70 000) and "src" which is associated with the ability of the virus to transform cells in culture and appears to code for a $6 \times 10^4$ daltons protein (8). It should be noted that the existence of a fifth gene is also postulated but the function of the hypothetical protein encoded in this part of the genome, adjacent to the poly A segment, is completely unclear. In order to understand the mechanism by which the synthesis of the different viral proteins is regulated, several attempts have been performed to characterize the viral specific polypeptides made in infected cells. Identification of the virus specific proteins is different in these cells because host protein synthesis is not inhibited by RNA tumor viruses and because the amounts synthesized are small relative to those proteins synthesized by the host. Some success have however been obtained using antibodies directed against detergent disrupted viruses or purified viral proteins.
Recent studies have indeed shown that synthesis of mature avian proteins in cell-infected with avian or mammalian viruses generally involves the cleavage of higher molecular weight precursors. This is namely true for the gs proteins which are synthesized from a common precursor which corresponds to the translation of the complete gag gene of the viral 30-40S RNA.
In avian viruses this precursor which is designated Pr 76 (Pr stands for precursor followed by the apparent molecular weight $\times 10^{-3}$ of the polypeptide) is cleaved to give rise to the structural proteins p27, p19, p15 and p12 of the virus. Either using the pactamycin technique (9) and the peptide mapping of the several intermediates in the processing of the 76 000 daltons precursor, it has been possible to deduce, in avian virus RSV, the order of the individual polypeptides p27, p19, p15 and p12 in the Pr 76 molecule as well as the possible scheme of its processing (9).
In cells infected either with avian, murine or bovine oncornaviruses a viral specific polypeptide much larger than the gs precursor can be detected. This 180-200 000 daltons protein is synthesized in a considerably lower (approximately 1/20 th) amount than the gag gene product. This large polypeptide can be precipitated by antibodies directed against the gs proteins but also by antibodies directed against the reverse transcriptase (10). It is now admitted that this Pr 180-200 gag-pol is the initial translation product that leads to the formation of mature reverse transcriptase. The Pr 180-200 gag-pol probably results from an occasional read-through of a single termination codon at the end of the gag gene. The low frequency (5%) of this read-through may account for the low amount of reverse transcriptase in the virion (11).
It seems clear that the viral genomic 30-40S RNA as such is not the messenger for the major envelope glycoproteins. In fact, it has been shown that it is a 20S RNA, only present in infected cells, which is the mRNA for these proteins. This 20S mRNA probably results from a

nuclear conversion of 35-40S RNA (12). In avian viruses, the primary translation product of this 20S RNA is a 7 x $10^5$ daltons non glycosylated polypeptide which is then partially glycosylated (13). (This raises the apparent molecular weight on acrylamide gels to 90-95 000). The two major glycoproteins are generated by proteolytic cleavage and addition of fucose residues to the carbohydrate side chains of this precursor (14).
In avian sarcoma viruses the putative src protein(s) involved in transformation is (are) not translated directly from the 30-40S RNA. The real messenger RNA for this (these) protein(s) is (are) not yet clearly defined (15, 16).

## II. Processing of the gag gene polypeptides

Relatively little is known about the enzymes involved in the cleavage of the different viral protein precursors. So far, the processing of the gag gene product has been one of the most extensively studied. In this part of the paper we shall deal with data which permit the partial identification of the enzymatic activities responsible for the processing of this gag precursor. It has been demonstrated that the viral 30-40S RNA can serve as a messenger for the gag polypeptide in an in vitro protein synthesizing system derived from Krebs ascites cells (17). If the viral RNA alone is added to this translation system, one observes the synthesis of a 76 000 daltons protein which can be precipitated by anti-gs antibodies. No mature gs proteins are detectable.
Von der Helm has shown that the addition of partially purified p15 protein (isolated from the virion) to a Krebs ascites lysate programmed with Rous sarcoma virus (RSV) RNA allows a fairly good processing of the 76 000 daltons polypeptide to occur (18). In this case p27 and probably p15 and p12 are produced but p19 is not observed. None of the other gs proteins p27, p19 or p12 does seem to have any proteolytic activity on the Pr 76. It is interesting to note that whilst partially disrupted RSV virus added to the in vitro system also induced the cleavage of the homologous gs precursor, partially disrupted Rauscher leukemia virus (of murine origin) does not. This proteolytic activity appears thus specific. As the Pr 76 processing only occurs when protein p15 is added to the in vitro translation system, Von der Helm concluded that the p15 protein cannot be generated autocatalytically from the Pr 76. For this reason, this author suggested that productive viral infection in host cell requires p15 in addition to RNA and reverse transcriptase.
The results we have obtained using an other translation system strongly suggest it not to be the case. Indeed, instead of using an in vitro protein synthesizing system, we injected the viral RNA from AMV viruses into frog oocytes. This "oocyte system" has proven to be very efficient in translating mRNA from very different origins as well as being capable of performing postranslational processing.
In frog oocytes, both 70S and 30-40S RNA from avian myeloblastosis virus are faithfully translated into the gag Pr 76 precursor. Moreover one observes the complete processing of this Pr 76 polypeptide (19). The result of a "pulse-chase" experiment performed both in AMV

infected fibroblasts or in oocytes injected with purified 30-40S RNA is presented on Fig. 2.

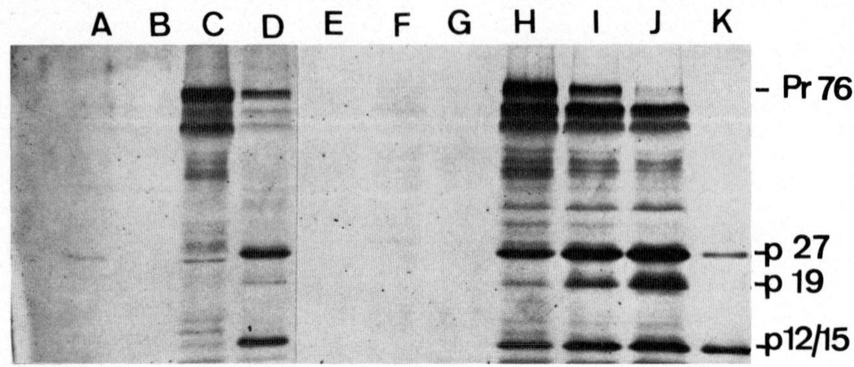


Fig. 1
Fluorograph of $NaDoDSO_4$ gel electrophoresis of immunoprecipitates of chick embryo fibroblasts or oocytes incubated with $^{35}S$ methionine. Immunoprecipitation was done with polyvalent antiserum against AMV. (A) Uninfected chick embryo fibroblasts pulse-labelled for 5 min; (B) as (A) but chased for 60 min; (C) AMV infected fibroblasts pulse labelled for 5 min; (D) as (C) but chased for 60 min; (E) non injected oocytes labelled for 20 hr; (F) and (G) as (E) but chased for 26 hr and 78 hr; (H) oocytes injected with 30-40S AMV RNA and labelled for 20 hr; (I) and (J) the same as (H) but chased for 24 hr and 78 hr; (K) $^{35}S$ methionine-labelled AMV marker.

It is clear from the data that all the final products of the processing of the gag protein Pr 76 as well as (most if not) all the cleavage intermediate polypeptides are made in the "oocyte" as they are in the infected cells. Even the p19 protein which is not made in the cell free translation systems is here clearly produced.
In order to learn more about the origin (viral or cellular) of the enzymatic activities involved in the cleavage of the Pr 76, we then performed the following experiment. Oocytes are injected with 30-40S RNA from AMV. They are subsequently incubated with $^{35}S$ methionine for a short period of time. In these conditions enough labelled Pr 76 is synthesized while no significant processing of this polypeptide has time to occur. After this incubation, the oocytes are homogenized and the extract is dialyzed in order to eliminate the free labelled amino-acid. Part of this homogenate is then injected in two separate batches of oocytes. The oocytes from one of these batches had received an injection of AMV-RNA 24 hours before and had been incubated without labelled amino-acid for this period of time. The other batch is constituted of control uninjected oocytes also incubated for 24 hours without labelled amino acids.

After injection of the extract containing the labelled Pr 76 the two batches of oocytes are incubated for another 48 hours in an unlabelled medium. At the end of this period, the oocytes are homogenized and their content of labelled viral proteins analyzed on acrylamide gels after immunoprecipitation.
The result of this experiment is given on Fig. 3. It is clear that the processing of the Pr 76 gag polypeptide is by far more pronounced in the oocytes which were allowed to synthesize viral proteins before the injection of the extract containing the labelled Pr 76.
It should be noted that the slight cleavage observed in non preinjected oocytes is quite normal since in the oocytes, the Pr 76 can be cleaved without any injection of other proteins.

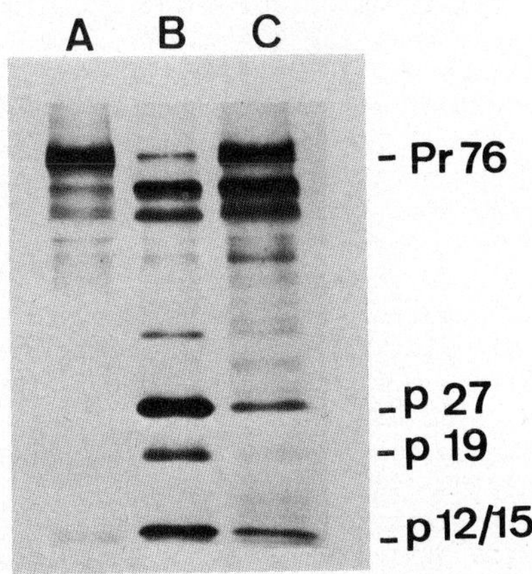


Fig. 2
Fluorograph of NaDoDSO$_4$ gel electrophoresis of immunoprecipitates of viral proteins made in oocytes. Immunoprecipitation was done with polyvalent antiserum against AMV.
(A) Oocytes injected with AMV RNA and incubated for 4 hr in an unlabelled medium.
In (B) the oocytes were preinjected with viral RNA 24 hr before injection of the cellular extract.
In (C) control oocytes not injected with viral RNA (for details see text).

## III. Discussion

From the results presented above, it can be concluded that a virus encoded protein is involved in the cleavage of the gag polypeptide. Whether this viral polypeptide is itself the specific processing protease or whether it activates (or modifies) a cellular enzyme present in the host infected cell is still unknown. It is also not clear whether this viral polypeptide is identical to one of the

structural proteins of the virion. There are however some arguments in favour of this hypothesis.
One comes from the fact that the protease activity which process the RSV gag polypeptide in vitro copurifies with the p15 structural protein from the virion (both in SDS gels and in a sephadex molecular filtration in the presence of guanidine-HCl). These observations contrast with the finding of Yoshinaka et al. who demonstrate (20) that a murine leukemia tumor virus (RLV) gag protein Pr 70 can be processed <u>in vitro</u> by a protein from the virion particle which does not copurify with any of the major gs protein.
A second argument is based on the following observations. Cells infected with MC 29 virus (21) or fibroblasts carrying the RAV-O endogenous virus (22) do not produce viral particles. One can however detect in both types of cells the synthesis of a 110-120 000 daltons polypeptide carrying two of the four gs antigens corresponding to proteins p27 and p19. This large polypeptide is the translation product of the viral genome RNA which is deleted in a region of the molecule including that coding for the p15 protein. As this 110-120 000 daltons polypeptide is not processed, a correlation has been suggested between the absence of p15 in this precursor and the lack of its processing (22).
The frog oocytes experiments described above clearly demonstrate that there is no need for the injection of a viral processing factor (VPF) together with the viral RNA to obtain a perfect cleavage of the gag polypeptide. Obviously, this VPF is synthesized in the oocyte from the viral RNA.
The question remains to know how this polypeptide is generated. The most amenable hypothesis would be that this factor is synthesized as part of the gag polypeptide. (This is obviously the case if this VPF coïncides with the p15). Then there are two possibilities which at the time being are equally probable.
One would be that this processing factor is autocatalytically cleaved from the gag precursor. Such a mechanism do exist in phage T4 infection. It has been demonstrated that the T4 PPase (a specific processing enzyme) is derived in this way from the P21 polypeptide (the precursor polypeptide made on gene 21) (23). The results obtained in the oocytes are compatible with such a mechanism.
The other possibility is that the cleavage of the viral processing factor from the gag precursor is mediated by a cellular enzyme. This hypothesis is supported by the findings of Eisenman et al. (24) showing that in non producing Rous sarcoma virus transformed hamster cells, the Pr 76 gag polypeptide is synthesized but not processed. The fusion of these infected hamster cells with permissive chick cells allows the Pr 76 to be cleaved and the virus to be rescued as in normal infection. This suggests that a cellular factor present in chick cells is required to some extent for the processing of the gag precursor. This cellular factor could be the enzyme which is needed to split the VPF from the gag precursor.
Should this be true this enzyme would be present in the frog oocytes. This is perhaps not surprising if one takes into account the fact that the oocyte is a typically undifferentiated cell and may contain enzymes which are no longer produced in differentiated cells. If such

an enzyme does exist in frog oocytes it does not show any specificity for the gag polypeptide of the avian viruses. We have indeed been able to translate the genomic viral RNA from the bovine leukemia virus in frog oocytes. Here again the gag polypeptide(s) (are) is produced and processed as in normal infected cells (25). Frog oocytes thus seem to be a promising tool for a further characterization of the enzymes involved in viral proteins processing.

## IV. Acknowledgements

This work was made possible through the financial support of the Fonds Cancérologique de la Caisse Générale d'Epargne et de Retraite and the State Contract "Actions concertées". J.G. is "Assistant au Fonds Cancérologique de la Caisse Générale d'Epargne et de Retraite", R.K. and D.P. are "Aspirants du Fonds National de la Recherche Scientifique", G.M. and G.H. are "Chercheurs Qualifiés du Fonds National de la Recherche Scientifique". M.T. held a Travel Fellowship awarded by the International Agency for Research on Cancer.

## V. References

1. Temin, H. and Mizutani, S., RNA-dependent DNA polymerase in virions of Rous sarcoma virus, Nature 226, 1211 (1970).
2. August, J., Bolognesi, D., Fleissner, E., Gilden, R. and Nowinski, R., A proposed nomenclature for the virion proteins of oncogenic RNA viruses, Virology 60, 595, 1974.
3. Ikeda, H., Hardy, W., Tress, E. and Fleissner, E., Chromatographic separation and antigenic analysis of proteins of the oncornaviruses, J. Virol. 16, 53 (1975).
4. Burny, A., Bex, F., Bruck, C., Cleuter, Y., Dekegel, D., Ghysdael, J., Kettmann, R., Leclercq, M., Mammerickx, R. and Portetelle, D., Biochemical studies on enzootic and sporadic types of bovine leucosis. In Anti-viral Mechanisms in the Control of Neoplasia (P. Chandra, ed) Plenum Press, New York (1978) in press.
5. Stromberg, K., Hurley, N., Davis, N., Rueckert, R. and Fleissner, R., Structural studies of avian myeloblastosis virus : comparison of polypeptide in virion and core component by dodecylsulfate polyacrylamide gel electrophoresis, J. Virol. 13, 513 (1974).
6. Duesberg, P., Physical properties of Rous sarcoma virus RNA, Proc. Natl. Acad. Sci. USA 60, 1511 (1968).
7. Furuichi, Y., Shatkin, A., Stavnezer, E. and Bishop, J., Blocked methylated 5'-terminal sequence in avian sarcoma virus RNA, Nature 257, 618 (1975).
8. Purchio, A., Erikson, E. and Erikson, R., Translation of 35S and subgenomic regions of avian sarcoma virus RNA, Proc. Natl. Acad. Sci. USA 74, 4661 (1977).
9. Vogt, V., Eisenman, R. and Diggelmann, H., Generation of avian myeloblastosis virus structural proteins by proteolytic cleavage of a precursor polypeptide, J. Mol. Biol. 96, 471 (1975).
10. Jamjoom, G., Naso, R. and Arlinghaus, R., Further characterization of intracellular precursor polyproteins of Rauscher leukemia virus, Virology 78, 11 (1977).

11. Philipson, L., Andersson, P., Oishevsky, U., Weinberg, R. and Baltimore, D., Translation of MuLV and MSV RNAs in nuclease-treated reticulocyte extracts : enhancement of the gag-pol polypeptide with yeast suppressor tRNA, Cell 13, 189 (1978).
12. Stacey, D. and Hanafusa, H., Nuclear conversion of microinjected avian leukosis virion RNA into an envelope-glycoprotein messenger, Nature 273, 779 (1978).
13. Shapiro, S., Strand, M. and August, J., High molecular weight precursor polypeptides to structural proteins of Rauscher murine leukemia virus, J. Mol. Biol. 107, 459 (1976).
14. England, J., Bolognesi, D.P., Dietschold, B. and Halpern, M., Evidence that the precursor glycoprotein is cleaved to yield the major glycoprotein of avian tumor viruses, J. Virol. 21, 810 (1977).
15. Kamine, J., Burr, J. and Buchanan, J., Multiple forms of src gene proteins from Rous sarcoma virus RNA, Proc. Natl. Acad. Sci. USA 75, 366 (1978).
16. Purchio, A.F., Erikson, E., Brugge, J.S. and Erikson, R.L., Identification of a polypeptide encoded by the avian sarcoma virus src gene, Proc. Natl. Acad. Sci. USA 75, 1567 (1978).
17. Von der Helm K. and Duesberg, P.H., Translation of Rous sarcoma virus RNA in a cell-free system from ascites Krebs II cells, Proc. Natl. Acad. Sci. USA 72, 614 (1975).
18. Von der Helm, K., Cleavage of Rous sarcoma viral polypeptide precursor into internal structural proteins in vitro involves viral protein p16, Proc. Natl. Acad. Sci. USA 74, 911 (1977).
19. Ghysdael, J. Hubert, E., Travnicek, M., Bolognesi, D.P., Burny, A., Cleuter, Y., Huez, G., Kettmann, R., Marbaix, G., Portetelle, D. and Chantrenne, H., Frog oocytes synthesize and completely process the precursor polypeptide to virion structural proteins after microinjection of avian myeloblastosis virus RNA, Proc. Natl. Acad. Sci. USA 74, 3230 (1977).
20. Yoshinaka, Y. and Luftig, R.B., Properties of a Pr 70 proteolytic factor of murine leukemia viruses, Cell 12, 709 (1977).
21. Bister, K., Hayman, M.J. and Vogt, P.K., Defectiveness of avian myelocytomatosis virus MC 29 : isolation of long-term non producer cultures and analysis of virus-specific polypeptide synthesis, Virology 82, 431 (1977).
22. Eisenman, R., Shaikh, R. and Mason, W.S., Identification of an avian oncornavirus polyprotein in uninfected chick cells, Cell 14, 89 (1978).
23. Showe, M.K., Isobe, E. and Onorato, L., Bacteriophage $T_4$ prehead proteinase, J. Mol. Biol. 107, 55 (1976).
24. Eisenman, R., Vogt, V.M. and Diggelmann, H., Synthesis of avian RNA tumor virus structural proteins, Cold Spring Harb. Symp. Quant. Biol. 39, 1067 (1974).
25. Ghysdael, J., Hubert, E. and Cleuter, Y., Biosynthesis of bovine leukemia virus (BLV) major internal protein (p24) in infected cells and X. laevis oocytes microinjected with BLV 60-70S RNA, Arch. Internat. Physiol. Biochem. 85, 978-979 (1977).

# SYNTHESIS AND PROCESSING OF MULTICOMPONENT PROTEINS

J. R. Tata
National Institute for Medical Research
Mill Hill, London NW7 1AA, U.K.

## ABSTRACT

It is now becoming increasingly clear that a number of proteins, some with very different properties, are synthesized as common multicomponent precursors which are then cleaved. The best documented examples are the tumour virus proteins known as "gag" gene proteins, the pituitary polypeptide hormones ACTH, LPH, MSH synthesized as a common precursor with endorphin, and vitellogenin which is the precursor for egg yolk proteins phosvitin and lipovitellin. The precursor may be proteolytically cleaved in the same cell in which it is synthesized or in a different cell. Translation of precursor messenger RNA in heterologous cell-free systems has helped establish the multicomponent nature of such precursors as well as the order in which the component proteins are arranged. Processing (glycosylation, phosphorylation, lipidation and proteolysis) has been largely studied *in vivo* or by microinjection of precursor messenger into frog oocytes. The wider implications of synthesis and processing of multicomponent proteins are briefly discussed.

## INTRODUCTION

Recent work on several proteins in diverse biological systems has revealed that these are synthesized as part of a multicomponent precursor and then cleaved. It is appropriate to first define what is meant by a "multicomponent precursor" and to especially distinguish it from already well-known examples of multi-enzyme complexes.

Perhaps the best known examples of multi-enzyme complexes are enzymes with interrelated functions, such as the fatty acid synthetase and the aspartate transcarbamylase-carbamyl phosphate synthetase-dihydroorotase (ATC/CPS/DHO) complexes (Ref. 1). In these multi-enzyme systems, one is dealing with a single large polypeptide within which functionally different components are all active while still in a covalent linkage, i.e. the different activities do not have to be split off from the multi-enzyme unit. The examples to be discussed below are distinguishable from such multifunctional proteins by virtue of the fact that the individual components are cleaved off. These, in turn, could be divided into two categories: a) those that are proteolytically cleaved in the same cell in which they are synthesized, and b) those that are cleaved into their constituents in a different cell or tissue from the one in which they are synthesized and chemically modified.

## "GAG" GENE PROTEINS

A group of four proteins of avian sarcoma-leukosis virus produced in the host cell cytoplasm are known to be coded for by a single gene known as the "gag" gene (Refs. 2,3). Evidence from many laboratories has now confirmed that these proteins of molecular weights of 12,000, 15,000, 19,000 and 27,000 daltons are synthesized as a precursor of molecular weight 76,000 (Pr 76) on a single messenger RNA (situated at the 5' end of the viral genome) corresponding in size to that of the polypeptide. This conclusion has been arrived at by studying mRNA translation *in vivo*, in cell-free systems and after injection into *Xenopus* oocytes. The cleavage of the Pr 76 protein occurs soon after its synthesis in a stepwise manner but the exact sequence of cleavage has not yet been definitively established (see Fig. 1).

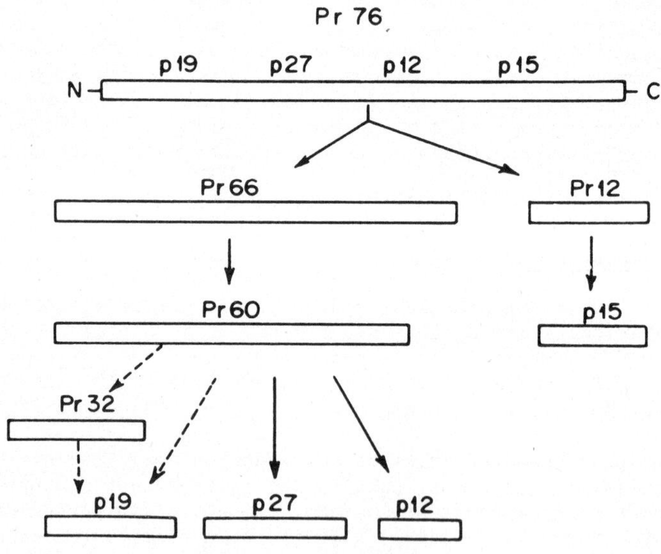


Fig. 1. Scheme showing the cleavage of Pr 76 into "gag" gene proteins of avian sarcoma-leukosis virus (from Ref. 3).

Two of the "gag" polypeptides are known to be phosphorylated. However, the most interesting feature about this multicomponent protein is that the first fragment of 15,000 molecular weight that is cleaved off from the Pr 76 precursor is thought to be the protease which is itself responsible for further proteolytic cleavage of the molecule. This viral multicomponent system and its proteolytic processing is further discussed in this Symposium by Huez.

## VITELLOGENIN

The egg yolk proteins, phosvitin and lipovitellin, are known to be synthesized as a large phosphoglycolipoprotein precursor of a subunit m.w. of 210,000 - 240,000 called vitellogenin (see Refs. 4,5 for reviews). Vitellogenin, which has almost identical composition in all egg-laying animals, is synthesized in the liver in vertebrates or in the fat body in insects; it is secreted into the blood or haemolymph and then cleaved into its components in oocytes or the ovary. The two major products of vitellogenin found in the yolk are remarkably different in their composition (see Fig. 2). For example, the two residues of phosvitin (molecular weight of around 32,000) for each vitellogenin molecule, have about 56% of amino acid residues as serine, most of which are phosphorylated. Phosvitin is lipid-free, has only one methionine residue, and is heavily glycosylated. On the other hand, the other component of vitellogenin, lipovitellin

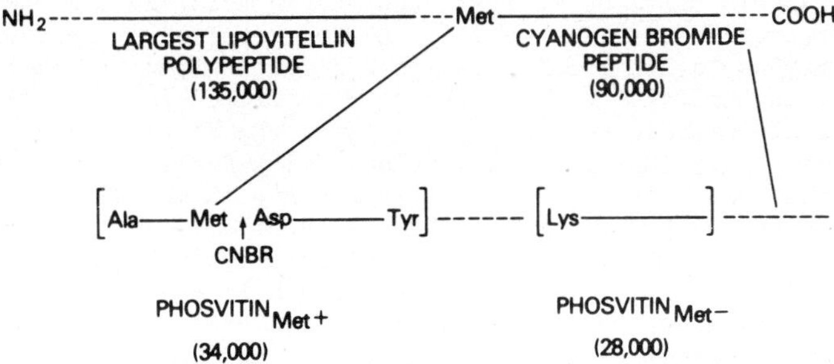


Fig. 2. The organization of two phosvitin and one lipovitellin residues in chicken vitellogenin as revealed by cell-free translation of messenger RNA (from Ref. 6).

(m.w. of about 120,000), has almost all the methionine residues of vitellogenin, not more than 4% serine, contains virtually no carbohydrate but is 20% lipid by weight. The major significance of this remarkable difference between the two egg yolk proteins is that different regions of the precursor vitellogenin polypeptide are modified or processed in entirely different ways (i.e. phosphorylated, glycosylated and lipidated).

The synthesis of vitellogenin in hepatocytes of vertebrates or fat body of insects in under obligatory hormonal control - estrogen in vertebrates and juvenile hormone in most insects. Administration of estrogen will induce the synthesis and secretion of vitellogenin equally well in liver of male as well as female egg-laying vertebrates. The induction of hepatocytes from males, which would normally never synthesize the protein, combined with the large size of the messenger RNA coding for the precursor ($2.0 - 2.5 \times 10^6$ daltons) has made it relatively easy to purify vitellogenin mRNA. One of the useful applications of purifying vitellogenin messenger is that an analysis of its translation in cell-free systems has made it possible to determine the organization of the cleavage products in the precursor. By combining peptide mapping, kinetics of incorporation of serine vs. methionine in the polypeptide synthesized from vitellogenin mRNA by a wheat germ cell-free system, Gordon *et al*. (Ref. 6) were able to derive the structure of chicken vitellogenin as depicted in Fig. 2. This approach confirmed the presence of two phosvitin residues in vitellogenin and demonstrated for the first time that both of these were grouped together at the carboxyl terminus of vitellogenin. Interestingly, they also found that only one of the two phosvitin molecules had a methionine residue.

Although relatively little is known about the chemical modifications of vitellogenin *in vivo*, it seems that these occur during or soon after its translation (see Ref. 4). The processes of phosphorylation, glycosylation and lipidation are all thought to occur while the still uncompleted or completed polypeptide chain is attached to the rough endoplasmic reticulum, vesicularized first in smooth endoplasmic reticulum and then within Golgi bodies, prior to its secretion as shown in Fig. 3. It is of some interest that all the enzymes known to be involved with the above modifications are associated with membranes of the endoplasmic reticulum or the Golgi apparatus (see Ref. 7). Since vitellogenin is a secreted protein, it would be expected to be synthesized on membrane-bound ribosomes. However, the localization of many modification enzymes in membranes of cytoplasmic organelles gives an added significance to the role of intracellular membranes in protein synthesis, especially where it involves the modification of nascent polypeptide chains before their completion.

Once secreted into the circulation, the fully modified vitellogenin dimer is taken up by the oocyte or the immature egg in which it is

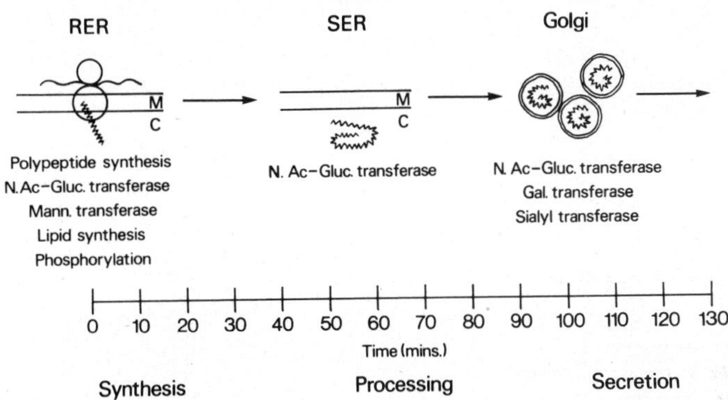


Fig. 3. Schematic representation of sequential modifications and secretion of *Xenopus* vitellogenin in the endoplasmic reticulum and Golgi apparatus before its secretion from the liver (from Ref. 5).

cleaved. In *Xenopus* oocytes the uptake of vitellogenin is thought to be governed by a process of endopinocytosis (Ref. 8) as determined by experiments in which oocytes were incubated with vitellogenin labelled in the intact animal. If the same labelled vitellogenin were to be microinjected into oocytes the protein is rapidly degraded rather than correctly cleaved and its cleavage products inserted into the crystalline yolk platelets. This finding suggests that the manner in which the circulating vitellogenin is "packaged" and taken up into oocytes must somehow protect it from lysosomal degradative enzymes. Berridge and Lane (Ref. 9) also observed the same phenomenon with *in vivo* synthesized vitellogenin injected into oocytes. However, they found that if vitellogenin were synthesized within the oocyte (which it normally does not do) by injecting it with mRNA extracted from the liver, the protein is correctly modified (at least as far as phosphorylation is concerned), cleaved and the resulting phosvitin and lipovitellin deposited into yolk platelets. Little is known about the proteolytic enzymes involved in splitting the vitellogenin molecule after its uptake by the oocyte, although Bergink and Wallace (Ref. 10) described a serine protease in *Xenopus* ovary which efficiently hydrolysed vitellogenin into phosvitin and lipovitellin. A few preliminary results also suggest that in *Xenopus* the proteolytic activity may be associated

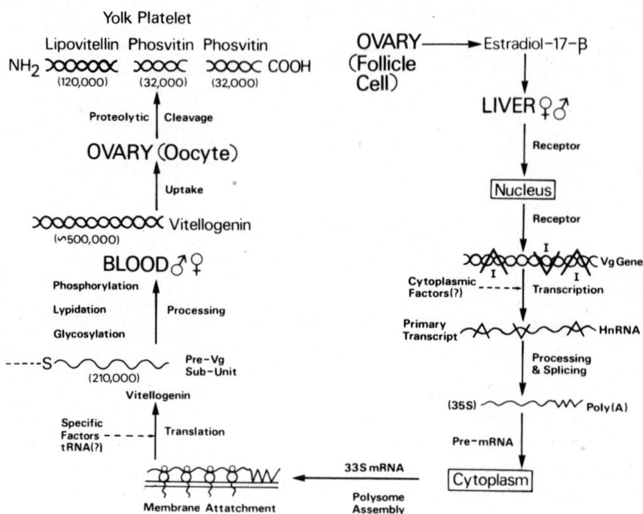


Fig. 4. Summary of events leading to the induction by estrogen of vitellogenin mRNA, its translation, chemical modification, secretion into blood, the uptake of the precursor into oocytes and its cleavage there into phosvitin and lipovitellin (based on Ref. 5).

with the yolk platelets themselves. Figure 4 summarizes the sequence of hormonal induction, translation, post-translation modifications, secretion and cleavage of vitellogenin in vertebrates.

## ACTH/LPH/ENDORPHIN PRECURSOR

Much interest has been recently generated around the opiate peptide endorphin, and its related derivative enkephalin, because of their important central nervous functions. It is now known that β-endorphin is covalently associated with the pituitary polypeptide hormones ACTH and β-LPH. Of even greater interest is the discovery that all these peptides are synthesized as a single component in the pituitary gland and then processed (glycosylated) and cleaved in a

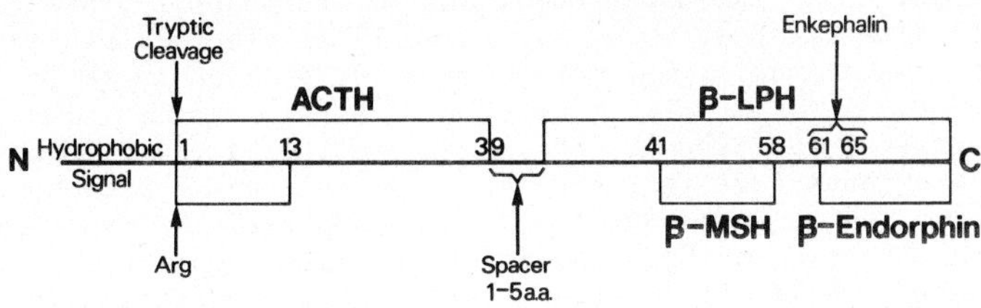


Fig. 5. The organization of ACTH/LPH precursor synthesized in the pituitary and the different cleavage sites to yield ACTH, β-LPH, β-MSH, β-endorphin and enkephalin (based on Ref. 11).

variety of ways (Refs. 11,12). Based on the cell-free translation of an mRNA which coded for a large peptide containing ACTH, Roberts and Herbert (Ref. 12) were able to assign the organization of the multicomponent precursor as depicted in Fig. 5. Of particular interest is the fact that this precursor contains within it a number of polypeptides with quite different physiological activities.

Relatively little is known as yet about the mode of processing and cleavage of the ACTH/LPH/endorphin precursor. However, it is clear from Fig. 5 that different permutations of cleavages at different sites would generate different products. ACTH is also known to be a glycoprotein and it would be of some interest to know how the precursor and its proteolytic cleavage products are post-translationally modified. In a recent study based on double-labelling and characterization of glycopeptides formed in mouse pituitary tumour cells capable of synthesizing the precursor, Roberts *et al.* (Ref. 13) were able to show that glycosylation and cleavage occur in a complex manner. They found that the precursor polypeptide initially acquires the core carbohydrate and this is followed by a series of cleavages accompanied with the addition of further core or peripheral sugars to the progressively smaller polypeptide fragments and the release of an N-terminal glycopeptide at each step.

## SIGNIFICANCE OF MULTICOMPONENT PROTEINS

The above examples suggest that multicomponent proteins may occur more widely than has been imagined until now and this raises the question of their significance and possible evolutionary advantage. The latter is easy to visualize for multi-enzyme systems such as fatty acid synthetase and ACT/CPS/DHO complexes. These systems must have arisen by gene fusion and the integration of units of a common function would contribute to a selective advantage for all the individual enzyme activities being covalently linked in a single large polypeptide. On the other hand, the multicomponent proteins that have been considered in this review are later cleaved into individual components which may or may not be directly related in their function.

An interrelated function of individual components would impart an evolutionary advantage in their being synthesized as a common precursor. Thus, an obvious advantage of the synthesis of the individual "gag" gene proteins as "Pr 76", or of phosvitin and lipovitellin as vitellogenin, would lie in these being formed simultaneously in a co-ordinate fashion. Perhaps their synthesis as a multicomponent precursor would also meet a requirement for their availability in fixed stoichiometric proportions. On these grounds, the significance of ACTH, β-LPH, β-MSH and β-endorphin (or enkephalin) being synthesized in the form of a common precursor is not readily clear at the present moment since each of these polypeptides is thought to have quite different physiological actions. One could however argue that we have not yet discovered all the multiple actions of these pituitary polypeptides, some of which may indeed be interrelated such that their formation has to be co-ordinated. Indeed, it has recently been suggested that ACTH may exert some important central nervous actions which are integrated with those of endorphin.

The above consideration of multicomponent proteins also raises some important biochemical and cell biological questions which have not yet been adequately investigated. For example, it would be important to know if there is any specificity in the proteolytic enzymes which cleave the multicomponent precursors, or is it simply that the pattern of cleavage is merely a function of the primary sequence of the substrate polypeptide? It is not unlikely that the oocyte which does not synthesize vitellogenin may be endowed with a protease which cleaves it highly efficiently into phosvitin and lipovitellin with little or no wastage. Similarly, it may be the presence of one or more special proteases in pituitary cells that make it possible to generate the various hormones and endorphin. In the case of the viral "gag" gene proteins, the highly interesting possibility has already been raised that one of these is itself a protease which specifically acts on the precursor. Another question that arises is how such multicomponent proteins which are "designed" to be readily cleaved are protected within the cell against a multitude of proteolytic enzymes. Segregation of proteins within intracellular membranous compartments would afford such a protection. For example, when the messenger coding for vitellogenin was injected into *Xenopus* oocytes, the nascent protein was found to be sequestered within a vesicular fraction which prevented its digestion by added trypsin (Ref. 14). Such a mechanism is most likely to be of general occurrence and would be of much interest in considering the question of movement within cells and the secretion of proteins.

Finally, it is important to realise that our knowledge of multicomponent protein systems described in this review has been gathered only relatively recently. It is also likely that many other similar multicomponent proteins have remained undetected or that the significance of larger precursors of proteins that are already known has not yet been fully recognised. For example, proinsulin could be considered as a multicomponent protein, especially if a function were found for the "C-peptide" which lies between the two insulin chains before cleavage. Whether or not this particular example is the most appropriate one for the present argument, it is not difficult to predict that many more examples of proteins being synthesized as multicomponent precursors will come to light in the next few years.

## REFERENCES


(1) G.R. Stark, Multifunctional proteins: one gene - more than one enzyme, Trends in Biochem. Sci. 2, 64,(1977).

(2) S.R. Weiss, H.E. Varmus and J.M. Bishop, The size and genetic composition of virus-specific RNAs in the cytoplasm of cells producing avian sarcoma-leukosis viruses, Cell 12, 983 (1977).

(3) R.N. Eisenman and V.M. Vogt, The biosynthesis of oncovirus proteins, Biochim. Biophys. Acta (Reviews on Cancer) 473, 187 (1978).

(4) M.J. Clemens, The regulation of egg yolk protein synthesis by steroid hormones, Progr. Biophys. Mol. Biol. 28, 69 (1974).

(5) J.R. Tata, The expression of the vitellogenin gene, Cell 9, 1 (1976).

(6) J.I. Gordon, R.G. Deeley, A.T.H. Burns, B.M. Patterson, J.L. Christmann and R.F. Goldberger, In vitro translation of avian vitellogenin messenger RNA, J. Biol. Chem. 252, 8320 (1977).

(7) G.C. Shore and J.R. Tata, Functions for polyribosome-membrane interactions in protein synthesis, Biochim. Biophys. Acta 472, 197 (1977).

(8) R.A. Wallace and D.W. Jared, Protein incorporation by isolated amphibian oocytes. V. Specificity for vitellogenin incorporation, J. Cell Biol. 69, 345 (1976).

(9) M.V. Berridge and C.D. Lane, Translation of Xenopus liver messenger RNA in Xenopus oocytes: Vitellogenin synthesis and conversion to yolk platelet proteins, Cell 8, 283 (1976).

(10) E.W. Bergink and R.A. Wallace, Precursor-product relationship between amphibian vitellogenin and phosvitin, J. Biol. Chem. 249, 2897 (1974).

(11) J.L. Roberts and E. Herbert, Characterization of a common precursor to corticotropin and β-lipotropin: Identification of β-lipotropin peptides and their arrangement relative to corticotropin in the precursor synthesized in a cell-free system, Proc. Natl. Acad. Sci. 74, 5300 (1977).

(12) J.L. Roberts and E. Herbert, Characterization of a common precursor to corticotropin and β-lipotropin: Cell-free synthesis of the precursor and identification of corticotropin peptides in the molecule, Proc. Natl. Acad. Sci. 74, 4826 (1977).

(13) J.L. Roberts, M. Phillips, P.A. Ross and E. Herbert, Steps involved in the processing of common precursor forms of adrenocorticotropin and β-endorphin in cultures of monkey pituitary cells, Biochemistry, in press (1978).

(14) T. Zehavi-Willner and C. Lane, Subcellular compartmentation of albumin and globin made in oocytes under the direction of injected messenger RNA, Cell 11, 683 (1977).

# THE MODE OF ANCHORING OF SUCRASE-ISOMALTASE TO THE SMALL INTESTINAL BRUSH-BORDER MEMBRANE AND ITS BIOSYNTHETIC IMPLICATIONS

G. Semenza, Laboratorium für Biochemie der ETH,
Universitätstrasse 16, CH 8092 Zürich, Switzerland

## ABSTRACT

Contrary to other intrinsic plasma membrane proteins, the sucrase-isomaltase complex is associated with the intestinal brush border membrane through a highly hydrophobic segment located not far from the N-terminal of one subunit (the isomaltase). This observation calls for additions and/or modifications to the generally accepted scheme of biosynthesis of other plasma membrane intrinsic proteins.

## INTRODUCTION

The two subunits of the small intestinal sucrase-isomaltase complex (a maltase-sucrase and a maltase-isomaltase), although not identical, have a considerable degree of homology (reviewed in refs. 1-3) and are apparently subjected to the same or to related biological control mechanism(s), as shown by the constancy of the sucrase-isomaltase ratios in random samples of human peroral biopsies (4), by their simultaneous (5,6) or almost simultaneous (7) appearance in development and their simultaneous absence in sucrose-isomaltase malabsorption (8,9), a mono-factorial genetic disease (10). Much is known, also, on their hormonal and dietary control and on the effects, sometimes apparently contradictory, of inhibitors of protein synthesis (reviewed in refs. 11-13). However, up to now virtually nothing is known as to the actual mechanism of biosynthesis of this membrane protein.

Clearly, any hypothesis on the biosynthesis of a membrane protein has to take into account the positioning of the protein in the membrane fabric. One of the best known membrane protein, glycophorin, spans the erythrocyte plasma membrane, with most of the protein including the N-terminal region and the sugar residues, protruding on the outside of the membrane. A hydrophobic segment occurs in the polypeptide chain not far from the C-terminal (14): it presumably interacts with the paraffinic chains of the bilayer lipids and helps in anchoring glycophorin to the fabric. The C-terminal region is located at the cellular side of the plasma membrane (15). Accordingly, the biosynthetic mechanism generally assumed for this kind of membrane proteins suggests that they are synthetized in the rough endoplastic reticulum, and that the nascent polypeptide chain enters "head on" (i.e., beginning with the N-terminal) the intracisternal space. The protein is not eventually released into the intracisternal space, but is kept in place, spanning across the membrane, by its hydrophobic area. (Glycosylation in the endoplastic reticulum is synchronized with the polypeptide biosynthesis (16)). Since the intracisternal space corresponds topologically to the extracellular space, the polypeptide chain is now in its final positioning (for a review, see ref. 17). A perhaps different mode of biosynthesis and assembly into the membrane could be envisaged in principle, for cytochrome $b_5$. This protein is reported to bind spontaneously to _performed_ lipid vesicles (18-21). It might, therefore, be synthetized like an ordinary, water soluble, protein and interact with the membrane _after_ complete biosynthesis.

In the case of small-intestinal sucrase-isomaltase complex, as it will be discussed in the following, neither of these two relatively straightforward biosynthetic mechanisms apply. As a matter of fact, even _a priori_ considerations lead to expect a more complicated biosynthetic mechanism: sucrase-isomaltase has a ml.wt. of about 220 000, i.e., several times larger than glycophorin or cytochrome $b_5$ and, contrary to these proteins, it is composed of two subunits.

The data which are summarized and discussed below deal with the positioning of sucrase-isomaltase in the small-intestinal brush border membrane. While many details are still missing, the broad features emerge rather clearly and allow some meaningful questions on the biosynthesis of this protein to be asked. It is quite possible that the conclusions reached and the questions put for sucrase-isomaltase may hold true for other large integral membrane proteins also, such as aminopeptidase, which has a similar size and possibly similar positioning in the membrane, at least as far as it can be judged from the data presently available (22). Sucrase-isomaltase - and probably the other digestive enzymes of the small intestine as well - is an intrinsic protein of the brush border membrane. Most of its mass, including its catalytic centers, carbohydrate chains, is located at the outer, luminal side of the membrane (23). No functional interaction can be detected between catalytic (glucosyl-hydrolytic) centers and the hydrophobic layer of the brush border membrane (2).

Sucrase-isomaltase (SI) can be detached from the membrane in one of at least two ways: by detergent solubilization (e.g. Triton X-100, product abbreviated T-SI), a process which presumably does not produce covalent changes in SI, or by controlled proteolytic digestion (papain-solubilized sucrase-isomaltase, abbreviated P-SI). As the solubilization by papain is irreversible, and P-SI, in contrast to T-SI, neither interacts with nascent liposomes (24) nor forms aggregates, papain must produce in still brush border-bound SI a chemical modification essential for its solubilization, i.e., cleave from the "body" of SI that part of the polypeptide chain which "anchors" it to the membrane.

## RESULTS AND DISCUSSION

It was thus deemed of interest to give a close scruting to the chemical differences between P-SI and T-SI. The first question was: is (are) the C-terminal areas(s) of sucrase and/or isomaltase involved in the cleavage by papain ? Table 1 compares the amino acids

Table 1 Amino acids released by carboxypeptidase Y from Triton-solubilized (T-SI) and papain-solubilized (P-SI) sucrase-isomaltase complex. The data are expressed as moles of amino acid released per mole of enzyme complex after digestion for 3 hrs. at 20$^{o}$C. (For details, see ref. 25).

| | P-SI | T-SI |
|---|---|---|
| Ser + Gln + Asn | 1.07 | 0.92 |
| Ile | 0.56 | 0.51 |
| Trp | 0.46 | 0.44 |
| Thr | 0.35 | 0.35 |
| Leu | 0.30 | 0.21 |

released by carboxypeptidase Y from P-SI and T-SI, respectively. Clearly, there are no significant differences. Since the amino acids released stemmed from both subunits (data not shown) (25) the identity in the amino acid patterns from P-SI and T-SI rules out that papain had cleaved off a part of the polypeptide chain in either subunit during the process of solubilization. Thus, whatever the mechanism of solubilisation by papain, it does not involve the C-terminal areas of either subunit.

Yet a cleavage of one or more peptide bonds does take place during papain solubilization (Fig. 1): the isomaltase band from T-SI has a smaller mobility in SDS-PAGE (and thus presumably a higher molecular weight) than its counterpart from P-SI. Thus, during papain solubilisation a decrease in apparent molecular weight of isomaltase occurs.

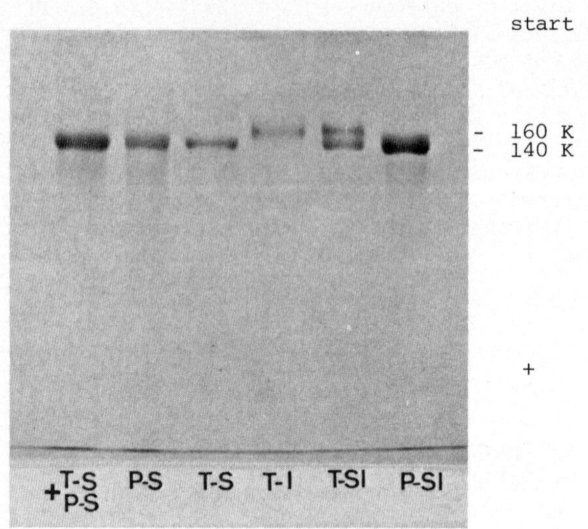


Figure 1 Slab SDS-PAGE of Triton-solubilized sucrase-isomaltase (T-SI), papain-solubilized sucrase-isomaltase (P-SI), isomaltase subunit from Triton-solubilized sucrase-isomaltase (T-I), sucrase subunit prepared from Triton-solubilized sucrase-isomaltase (T-S) and sucrase subunit prepared from papain-solubilized sucrase-isomaltase (P-S). (Ref. 25)

Table 2 N-Terminal amino acids of Triton-solubilized sucrase-isomaltase, papain-solubilized sucrase-isomaltase and individual subunits

The N-terminals were identified by dansylation. For details see ref. 25.

| | N-terminal(s) |
|---|---|
| Triton-solubilized sucrase-isomaltase | Ile, Ala |
| Papain-solubilized sucrase-isomaliase | Ile (Glu,Gly,Thr) |
| Isomaltase subunit from T-SI | Ala |
| Sucrase subunit from T-SI | Ile |
| Sucrase subunit from P-SI | Ile |

As the C-terminal areas are unaffected (see above) we expected the N-terminal of isomaltase to change during papain solubilization, and indeed this was found to be the case (Table 2). Thus, during papain solubilization, a segment of the polypeptide chain of the isomaltase subunit is cleaved off from its N-terminal end. Its cleavage leads to the irreversible solubilization of sucrase-isomaltase.

Sucrase, as judged from its unchanged apparent molecular weight, C- and N-terminals are unaffected by papain treatment.

Sequenation of the isomaltase subunit from T-SI yielded the following N-terminal sequence (ref. 26):

(CHO) [on residue 11, Thr]

Ala-Val-Asn-Ala-Phe-Ser-Gly-Leu-Glu-Ile-Thr-Leu-Ile-Val-

1 5 10

Leu-Phe-Val-Ile-Val-Phe-Ile-Ile-Ala-Ile-Ala-Leu-Ile-Ala

15 20 25

Val-Leu-Ala-x-x-x-Pro-Ala-Val

30 35

The occurrence of a highly hydrophobic area (12 to 31) agrees with the "anchoring area" of this intrinsic membrane protein being located not far from the N-terminal of the isomaltase subunit. Independent confirmation of the peripheral position of sucrase is also provided by its preferential solubilization by citraconylation (25) and by the e.m. observation that anti-isomaltase antibodies form precipitates closer to the apparent "unit membrane" than the anti-sucrase antibodies (27). It seems thus beyond doubt that sucrase-isomaltase is anchored to the brush border membrane via a hydrophobic segment located not far from the N-terminal (rather the C-terminal, as in most other intrinsic membrane proteins) of one subunit, i.e., the isomaltase subunit.

How does sucrase-isomaltase come to be positioned in this way ? It seems rather unlikely (although not entirely impossible) that it is synthetized as a water-soluble protein and is inserted into the plasma membrane afterwards, since sucrase-isomaltase, contrary to, e.g., cytochrome $b_5$ (18-21) does not associate with performed liposomes of either phosphatidylcholine or asolectin (24).

A more likely hypothesis is that the hydrophobic segment at the N-terminal area of isomaltase "sticks" in the hydrophobic layer of the endoplastic reticulum membrane during biosynthesis, or reassociate with it soon afterwards. As a matter of fact, it is possible that this N-terminal segment is a kind of "signal"peptide (28,29) (which, however, would not be cleaved off due to its "sticking" in the membrane, to lack of appropriate "signal peptidase", to its primary and/or secondary structure to other reasons).

The absence of a significant role of the C-terminal areas of either subunit in the anchoring of sucrase-isomaltase to the membrane is perhaps easier to explain: it is conceivable that freshly synthetized (and inserted) sucrase and/or isomaltase may possess (a) C-terminal segment(s) spanning across the membrane. Endopeptidases (e.g., of pancreatic origin) would cleave off the(se) segment(s) by acting from the lumen. Sucrase-isomaltase would not be lost into the lumen due to its hydrophobic anchor at the N-terminal of the isomaltase subunit.

An interesting and complicated problem is the coordination in biosynthesis and the mechanism of association of the two subunits. Clearly, since the interaction of sucrase is mediated (mainly or exclusively) by isomaltase, the cleavage of a (hypothetical) membrane anchor of sucrase at its C-terminal should be subsequent to its association with isomaltase.

A highly speculative hypothesis - whose only merit at the moment is that of explaining in a single framework the related biological control of sucrase and isomaltase, their association and their peculiar anchoring in the brush border membrane - would be the following. The homology between sucrase and isomaltase (1) indicates a common origin via gene duplication. If the two cistrons are still contiguous and code a two-cistron mRNA,a very high molecular weight polypeptide chain would be synthetized and inserted into the ER membrane, beginning with the N-terminal area of isomaltase (a modified "signal"?), the rest of the isomaltase, sucrase and finally the hypothetical additional anchor at the C-terminal of sucrase. After folding and glycosylation, proteolysis would remove the hypothetical C-terminal "anchor" and also split the single isomaltase-???-sucrase chain into the two "ripe" subunits. The subunits would be kept associated by (some of) the non-covalent interactions which had formed during the folding of the hypothetical single-chain precursor.

Naturally, this mechanism would be similar to that of chymotrypsinogen or proinsulin activation, and an enzymatically inactive membrane protein crossreacting with sucrase-isomaltase has indeed been reported and isolated from small intestine of baby rabbits who had not developed yet either sucrase or isomaltase activity (30). However, it should be pointed out clearly that the hypothesis of a single-chain (active or inactive) "precursor" has no direct experimental basis at the moment.

It is hoped that future work will answer some of the questions raised.

## REFERENCES


1. G. Semenza, (1976) Membranes and Diseases, L. Bolis & A. Leaf, Raven Press, New York. p. 243.

2. G. Semenza, (1976) The Enzymes of Biological Membranes, A. Martonosi, Plenum Press, New York, vol. 3, p. 349.

3. G. Semenza, (1978)"Oligosaccharidases and Disaccharidases of small intestinal brush borders" in Handbook Series in Nutrition and Food,M. Rechcigl, Jr., CRC Press Inc., Ohio, Section B, in press.

4. S.Auricchio, A. Rubino, R. Tosi, G. Semenza, M. Landolt, H. Kistler and A. Prader, Enzym. Biol. Clin. 3, 193 (1963).

5. A. Rubino, F. Zimbalatti and S. Auricchio, Biochim. Biophys. Acta 92, 305 (1964)

6. A. Dahlqvist and T. Lindberg, Clin. Sci. 30, 517 (1966).

7. K. Yamada, S. Moriuchi and N. Hosoya, J. Nutr. Sci.Vitaminol 24, 177 (1978).

8. S. Auricchio, A. Rubino, A. Prader, J. Rey, J. Jos, J. Frézal and D. Davidson, J. Pediatr. 66, 555 (1965).

9. G. Semenza, S. Auricchio, A. Rubino, A. Prader and J.D. Welsh, Biochim. Biophys. Acta, 195, 386 (1965).

10. K.R. Kerry and R.R.W. Townley, Aust. Paediatr. J. 1, 223 (1965).

11. Koldovský, O. (1972) Nutrition and Development J. Wiley & Sons, New York, p. 135.

12. K.M. Brown, J. Exptl. Zool.177, 493 (1971).

13. F. Moog, A.E. Denes and P.M. Powell, Developm. Biol. 35, 143 (1973).

14. H. Furthmayr, R.E. Galardy, M. Tomita and V.T. Marchesi, Arch. Biochem. Biophys. 185, 21 (1978).

15. S.F. Cotmore, H. Furthmayr and V.T. Marchesi, J. Mol. Biology 113, 539 (1977).

16. J.E. Rothman and H.F. Lodish, Nature, 269, 775 (1977).

17. D.D. Sabatini and G. Kreibich, The Enzymes of Biological Membranes (A. Martonosi, Plenum Press, New York) 2, 531 (1976).

18. L. Spatz and P. Strittmatter, Proc. Natl. Acad. Sci. 68, 1042, (1971).

19. L. Spatz and P. Strittmatter, J. Biol. Chem. 248, 793 (1973).

20. M.A. Roseman, P.W. Holloway, M.A. Calabro and T.E. Thompson, J. Biol. Chem. 252, 4842 (1977).

21. W.L.C. Vaz, H. Vogel, F. Jähnig and R.H. Austin, FEBS Letters, 87, 269 (1978).

22. D. Louvard, M. Semeriva and S. Maroux, J. Mol. Biol. 106, 1023, (1976).

23. G. Semenza, Handbook of Physiology (C.F. Code et al, eds.) V, 2543 (1968).

24. J. Brunner, H. Hauser & G. Semenza, J. Biol. Chem. (1978) in press.

25. J. Brunner, H. Hauser, H. Braun, K.J. Wilson, H. Wacker, B. O'Neill and G. Semenza, J. Biol. Chem. (1978) submitted for publication.

26. J. Brunner, G. Frank, H. Hauser, H. Wacker, T.J. Wilson, Semenza, G. and H. Zuber, FEBS Letters 1978.

27. Y. Nishi and Y. Takesue, Abstract, (32nd Annual Meeting of the Japanese Society of Electron Microscopy, Nagoya/Japan) 1976.

28. G. Blobel and B. Dobberstein, J. Cell Biol. 67, 835, (1975).

29. G. Blobel and B. Dobberstein, J. Cell Biol. 67, 851, (1975).

30. R. Dubs, R. Gitzelmann, B. Steinmann and J. Lindenmann, Helv. Paediatr. Acta 30, 89 (1975).

ACKNOWLEDGEMENTS:

Our work was partially supported by the SNSF, Berne, and Nestlé Alimentana SA., Vevey. To them thanks are due.

# THE SYNTHESIS AND PROCESSING OF COLLAGEN PRECURSORS

Michael E. Grant, J. Godfrey Heathcote and
Kathryn S.E. Cheah
Department of Medical Biochemistry, University of
Manchester Medical School, Manchester M13 9PT, U.K.

## 1. INTRODUCTION

Collagen is the most abundant protein in the vertebrate body, and it is the major constituent of most connective tissues. The characteristic differences in the properties of tissues such as tendon, cartilage, artery, skin, _etc_. are probably explained by the varying amounts of proteoglycans, glycoproteins and elastin associated with the collagen. However, in the last decade it has become clear that several genetically distinct forms of collagen can occur in the extracellular matrix. This new insight into the biology of connective tissues has provided a great stimulus to research into the chemistry of collagen and in this brief review the recent advances in our understanding of the mechanisms of assembly of these collagen types will be discussed (for more detailed reviews see 1 and 2).

## 2. THE NATURE OF COLLAGEN AND PROCOLLAGEN

The physical strength of collagen fibres is a consequence of the molecular structure of the collagen fibril monomer (tropocollagen), its packing arrangement in the collagen fibril, and the formation of covalent cross-links between the monomers in the fibrils. The collagen fibril monomer is essentially a stiff rod-like molecule (mol. wt. approx. 300 000) composed of three polypeptides, $\alpha$ chains, arranged in a triple-helical conformation. Early studies revealed that the tropocollagen readily isolated from tendon, skin and bone consisted of two identical chains ($\alpha 1$) and a distinct but homologous third chain ($\alpha 2$). From the work of Miller and others (for review see 3) it has become apparent that different interstitial tissues may contain one or more different types of fibrillar collagen (Table 1). Collagen types I, II and III are well documented (3) but the status of type IV, originally isolated by Kefalides (4) from lens capsule and glomerular basement membrane, is less clearly defined; and evidence for other molecular forms of collagen has appeared recently (5-7).

TABLE 1 Types of Collagen and Tissue Distribution

| Type | Molecular form | Tissue Distribution* | Distinctive features |
|---|---|---|---|
| I | $[\alpha 1(I)]_2\alpha 2$ | Bone, dermis, tendon, dentin, cornea, etc. | subunit heterogeneity, < 10 Hylys per chain low carbohydrate+ |
| II | $[\alpha 1(II)]_3$ | Cartilages, vitreous humour | > 10 Hylys per chain 4% carbohydrate |
| III | $[\alpha 1(III)]_3$ | Foetal and infant dermis, cardio-vascular system, uterine wall, etc. | cysteine, high Hypro, low carbohydrate |
| IV | $[\alpha 1(IV)]_3$ | Lens capsule, glomerular basement membrane | high 3-Hypro, high Hylys, low Ala, 12% carbohydrate |

* Tissues mentioned contain predominantly the collagen type indicated but many tissues have been found to contain more than one type.

\+ Carbohydrate occurs as Gal-Hylys and/or Glc-Gal-Hylys

To date the complete primary sequence of an α chain from a single species has not been determined but a composite sequence for the α1(I) chain is available from studies of residues 1-418 of rat skin α1(I) and residues 419-1052 of calf skin α1(I) (8). Over 95 per cent of the amino acids are arranged in triplets of structure -Gly-X-Y- and at both N- and C-termini are short non-triplet sequences (telopeptides) which are the remnants of longer sequences found in the collagen precursor polypeptides. In the repeating triplets the 'X-positions' and 'Y-positions' can be occupied by a variety of amino acids but frequently proline (Pro) is found in the X-position and hydroxyproline (Hypro) in the Y-position. Hydroxyproline and also hydroxylysine (Hylys) have been considered specific to collagen but they do occur in short collagen-like sequences in a few other proteins eg. C1q component of complement and acetylcholinesterase, and Hypro is a significant constituent of plant cell wall proteins. The roles of Hypro and Hylys in collagen molecules are discussed below but it should be noted that although the proportion of Hypro residues in an α chain is relatively constant (100-130 residues/1000 amino acids), the proportion of Hylys varies markedly from approx. 5 to 50 residues/1000 amino acids. The Hylys residues provide the sites for attachment of galactose and glucosylgalactose. Variations in the total hexose content and also in the ratio of monosaccharide to disaccharide units are found to occur both between collagens of different genetic types and within a single genetic type, particularly type I (1).

Under physiological conditions of pH, temperature and ionic strength, the collagen fibril monomer is insoluble and aggregates to form microfibrils. How cells can synthesise and secrete apparently insoluble molecules was not explained until it was demonstrated that collagen is initially synthesised as a larger, soluble triple-helical precursor, procollagen (Fig. 1). Most of the evidence for the existence of procollagen has come from experiments conducted

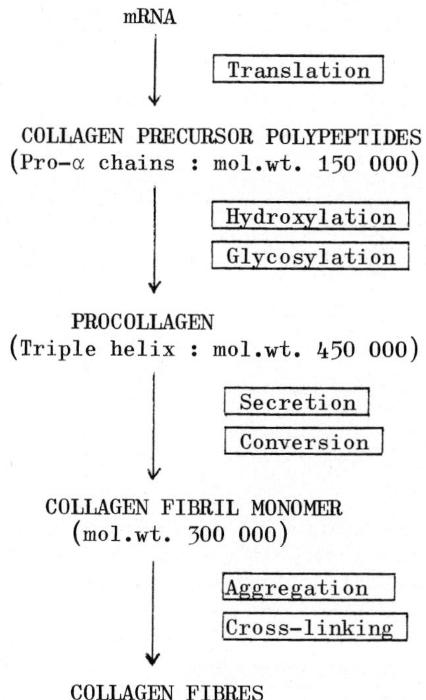


Fig. 1. Outline of collagen biosynthesis in interstitial connective tissues.

with cells synthesising type I collagen. Although type I procollagen is best characterised (Fig. 2), it is now apparent that all four collagen types are synthesised in precursor forms with non-helical peptide extensions at their N- and C-termini of approx. 15000 and 35000 respectively. Thus, the mol. wt. of the complete pro-α chains of collagen types I and II is approx. 150 000 (9), but the equivalent value for the precursor polypeptides of basement membrane collagen is approx. 180 000 (10) although the reason for this difference is not yet clear.

Analyses of the extension peptides indicate that their amino acid composition is quite unlike that of collagen and more typical of globular proteins (11,12); and particularly significant is their content of cysteine which is not found in the fibril monomer of collagen types I and II. Inter-chain disulphide bonds are located only in the C-terminal extensions of procollagen types I and II although intrachain disulphide bonding occurs in the N-terminal extensions. It is also of interest that the N-terminal extension contains three distinct structural regions. Sequence analysis of these peptides from pro-α1(I), pro-α2 and pro-α1(III) has indicated that a short collagenous region of 11 triplets of -Gly-X-Y- occurs in this essentially non-collagenous domain (Fig. 2).

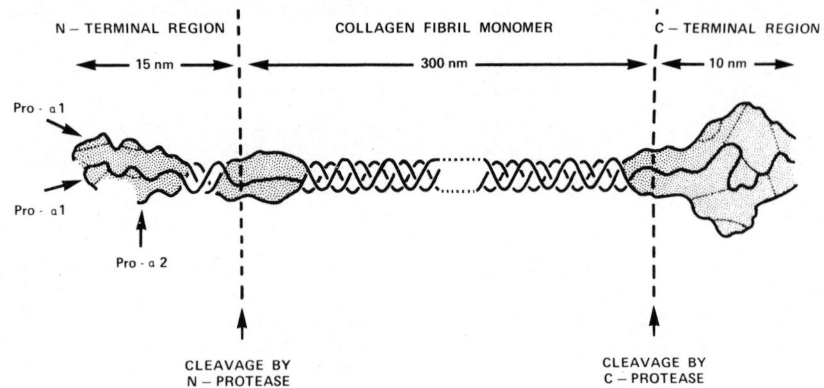


Fig. 2. Diagrammatic representation of the structure of type I procollagen indicating non-helical (shaded) and helical domains, and the location of disulphide bonds (- - -).

## 3. TRANSLATION OF PROCOLLAGEN mRNA

Considerable attention has been focussed on the mechanisms by which the cell can coordinate the translation of pro-α1(I) and pro-α2 chains so that the correct amounts of each polypeptide are produced for the assembly of type I procollagen. The possibility that collagen was derived from a large single chain precursor has been proposed but there is little evidence in favour of this concept since size determinations of collagen synthesising polysomes and pulse-chase experiments have all been consistent with the simultaneous translation of separate mRNAs for pro-α1(I) and pro-α2 (for review see 13).

It can be predicted that mRNA for pro-α chains must comprise at least 4500 nucleotides corresponding to a minimum mol. wt. of $1.5 \times 10^6$ and an expected sed. coeff. of 26-28S. PolyA-rich RNA extracts from rapidly developing connective tissues have been shown to be capable of directing the synthesis of collagenase-susceptible peptides in cell-free systems. Initial reports on the size of collagen mRNA were concerned with species sedimenting at 22-24S but subsequently species of 28-30S (mol. wt. 1.7-1.8 x $10^6$) have been isolated and shown to direct the synthesis of pro-α chains in a reticulocyte-lysate system. The yield of high mol. wt. translational products was low against an appreciable background level of protein synthesis endogenous to the cell-free system. Although greater success has been reported with wheat-germ cell-free systems incomplete procollagen chains were also synthesised. Indeed the wheat-germ system appears to favour the synthesis of proteins less than 100 000 mol. wt. (for detailed discussion see 13).

In more recent studies in our laboratory we have found the reticulocyte-lysate cell-free system pretreated to remove endogenous mRNA (14), to be very efficient in the synthesis of collagenous polypeptides when programmed with polyA-rich mRNA from chick tendon cells. Analysis of the translation products on SDS-polyacrylamide gels demonstrated the synthesis of high mol. wt. polypeptides which migrated in the region of standards of chick tendon pro-α1(I) and pro-α2 chains (Fig. 3). It should be noted however that unlike the procollagen polypeptides secreted by tendon cells the products of cell-free translation do not undergo the post-translational modifications of hydroxylation and glycosylation. Comparison of the translational products should therefore be made with unhydroxylated pro-α chains which have been shown to have a slightly higher mobility on SDS-polyacrylamide gels than the normal hydroxylated polypeptides (9). This observation is confirmed by the results in Fig. 3. Although the difference in mobility between band B (Fig. 3) and unhydroxylated pro-α2 is not very marked, it is of particular significance that the major high mol. wt. product, band A, has a mobility slower than unhydroxylated pro-α1(I). These bands have been shown to be collagenase-sensitive and appear to represent slightly higher mol. wt. precursor polypeptides. Such an observation is in accordance with the 'signal hypothesis' (15) which would predict that procollagen mRNA should code for pre-pro-α chains.

## 4. HYDROXYLATION AND GLYCOSYLATION REACTIONS

Hydroxyproline and hydroxylysine are not introduced into the molecule by the usual steps of polypeptide assembly but are formed by post-translational hydroxylation of peptide-bound Pro and Lys residues. The hydroxylation of Pro residues by proline hydroxylase gives rise predominantly to trans-4-hydroxyproline but all collagens contain some trans-3-hydroxyproline. Sequence analyses of vertebrate collagens indicate that 4-Hypro is found only in the Y-position of the typical collagen triplet -Gly-X-Y-. In contrast, 3-Hypro residues have been located only in the sequence -Gly- 3-Hypro - 4-Hypro - which occurs only once in the α1(I) chain of skin collagen but may occur several times in basement membrane collagens (1).

Prolyl hydroxylase has been isolated as a pure protein from several sources and extensively characterized (see 2). The enzyme does not hydroxylate free Pro and the minimum substrate requirement is an -X-Pro-Gly-triplet. Recently it has been possible to separate prolyl 3-hydroxylase and 4-hydroxylase activities and although

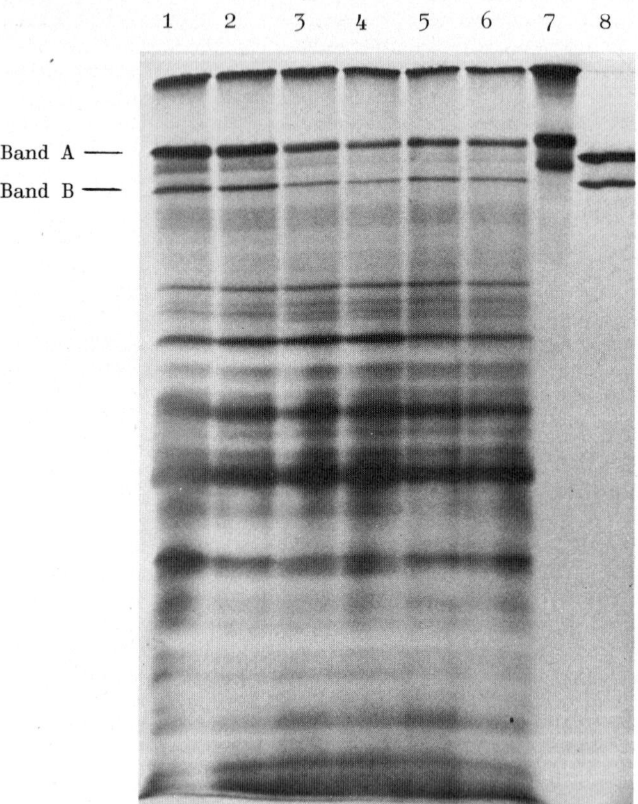


Fig. 3. Fluorogram of SDS-polyacrylamide gel of products of translation of a polyA-rich RNA from chick tendon cells. Slots 1 & 2, 3 & 4, 5 & 6 are duplicate samples of three different RNA preparations incubated with [$^{35}$S]Met in a messenger-dependent reticulocyte-lysate cell-free system. Slots 7 & 8 are [$^{14}$C]Pro-labelled standards of chick tendon pro-α chains and unhydroxylated pro-α chains, respectively, prepared according to the method of Harwood <u>et al</u>. (9).

these two enzymes share many common properties it appears that the main substrate for 3-Hypro synthesis is the sequence - Gly-Pro-4-Hypro-Gly- (16). Both prolyl hydroxylases belong to the group of

2-oxoglutarate dioxygenases which require molecular oxygen, ferrous iron, 2-oxoglutarate and a reducing agent (2). The 2-oxoglutarate is stoichiometrically decarboxylated during these reactions and one atom of the oxygen molecule is incorporated into the hydroxyl group while the other is incorporated into the succinate. The requirements for these cofactors and cosubstrates are specific although it has been reported that several reducing agents could substitute for ascorbate. However, recent kinetic studies with pure prolyl 4-hydroxylase indicates that the hydroxylation reaction was completely dependent on ascorbate (17).

The reaction in which peptidyl-lysine is hydroxylated during collagen biosynthesis is essentially the same as that described above for the hydroxylation of proline. A specific lysyl hydroxylase which also belongs to the enzyme group of 2-oxoglutarate dioxygenases (2) has been purified 300-fold from chick embryo cartilage extract (18) and although Hylys in collagen is usually found in Gly-X-Hylys-Gly- sequences, the purified enzyme will hydroxylate lysine residues in non-collagenous proteins *in vitro* (19). Thus, the substrate requirements for lysyl hydroxylase do not appear to be as rigid as for prolyl hydroxylase and this might account for the presence of Hylys residues in the non-helical telopeptide regions of collagen α-chains (see Section 6 below).

As noted above the carbohydrate residues found in extracellular collagen fibres are linked to Hylys and galactosyl-O-β-hydroxylysine and O-α-D-glucosyl(1-2)-O-β-D-galactosylhydroxylysine are now recognized as structural features common to all vertebrate collagens. Thus, collagen can be classified as a glycoprotein but it is noteworthy that in studies on the extension peptides of procollagen type I it has become evident that the N- and C-terminal regions contain sugars more typical of glycoproteins (12,20). In particular the C-terminal peptides contain concentrations of N-acetylglucosamine and mannose consistent with there being one or two oligosaccharide units per chain. The significance of these carbohydrate moieties in pro-α chains is not understood.

The synthesis of hydroxylysine glycosides involves two enzymes: collagen galactosyltransferase which transfers galactose from UDP-Gal to peptidyl-Hylys, and collagen glucosyltransferase which adds glucose from UDP-Glc to galactosyl-hydroxylysine. These enzymes require $Mn^{++}$ as a cofactor and their properties have been reviewed recently by Kivirikko & Myllylä (21). The role of these glycosides in collagen structure has not been defined but it seems likely that the sugars projecting out from the surface of the collagen fibril monomer may influence the size of collagen fibrils formed.

Studies on the subcellular location of prolyl hydroxylase and lysyl hydroxylase indicate these enzymes to be membrane-bound and associated almost exclusively with rough microsomal fractions (1,2). Such a location is consistent with analyses demonstrating that hydroxylation of Pro and Lys normally commences as the growing pro-α chains are assembled on large membrane-bound polyribosomes.

Similarly, the glycosyl transferases are bound to the membranes of the endoplasmic reticulum and both galactosylation and glucosylation are believed to begin at the level of the nascent peptides. Thus, the enzymes responsible for the secondary modifications of procollagen polypeptides may comprise a multi-enzyme system bound to the internal face of the cisternae of the rough ER.

Another important feature which all these enzymes have in common is the fact that they will not act on a helical substrate. Accordingly, the conformation of the substrate plays a critical role in controlling hydroxylation and glycosylation so that the degree of post-translational modification depends not only on the activities of the enzymes concerned but also on the rate of formation of the triple-helix (1,2).

## 5. ASSEMBLY AND SECRETION OF PROCOLLAGEN

The mechanisms by which the pro-α chains associate to form the triple-helix are unknown but the presence of covalent disulphide linkages between the C-terminal peptides suggests that the extension peptides may play a role in the assembly process. Determinations of the state of aggregation and conformation of procollagen polypeptides during biosynthesis have revealed a remarkable correlation between the time at which inter-chain bonds appear among pro-α chains and the time at which the protein becomes helical. Both events seem to occur in the rough ER after release of the pro-α chains from ribosomes. These investigations together with studies indicating that inter-chain disulphide bonds facilitate the renaturation of the triple-helix after thermal denaturation support the hypothesis that disulphide bonding must occur before triple-helix formation can occur at a biologically appropriate rate (1,2).

The rate of helix formation has been shown to vary from cell type to cell type and hence the time available for the hydroxylases and glycosyltransferases to act on the non-helical substrate can vary. The higher levels of Hylys and Hylys-glycosides in cartilage collagen and lens capsule collagen as compared to tendon collagen (see Table 1) can therefore be accounted for in part by the longer time required to achieve a helical conformation in chondrocytes and lens epithelial cells (1). It is not known whether disulphide bond formation is enzyme mediated but it is of interest that protein disulphide isomerase has been detected in tendon and cartilage cells (22) and the level of enzyme is lower in cartilage cells where the rate of disulphide bond formation between pro-α's is slower.

In experiments where Hypro and Hylys synthesis has been inhibited by removal of cofactors of the hydroxylation process it has been found that disulphide bonds can still be formed between pro-α chains (9). Under these conditions, however, the unhydroxylated procollagen (protocollagen) cannot form a triple-helix at 37°C for the presence of approx. 100 Hypro residues/pro-α chain is a critical requirement for helix stability under physiological conditions.

Thus, the hydroxylation of Pro is believed to play a fundamental role in maintaining collagen structure by forming additional intramolecular hydrogen bonds.through water molecules as intermediaries (2,23). When hydroxylation is inhibited, the unhydroxylated procollagen synthesised by tendon and cartilage cells is not secreted at a normal rate, suggesting that hydroxylation and the triple-helical conformation may be a requirement for optimal secretion. This observation is supported by other experiments in which proline analogues incorporated into pro-α chains have been shown to interfere with the folding of the helix and to lead to a decreased rate of secretion (2). No definitive data are available on the nature or mechanisms of the "conformation-dependent barrier" operating in cells secreting procollagen but it is clear from a variety of experimental approaches that procollagen follows the route of secretion taken by other extracellular proteins (1,2).

## 6. EXTRACELLULAR MODIFICATIONS

The first extracellular modification occurring in collagen synthesis is the conversion of soluble procollagen molecules to collagen fibril monomers which self-assemble into fibres displaying the characteristic 64 nm periodicity when viewed in the electron microscope. This conversion process is under the control of two extracellular proteases which remove the N- and C-terminal extensions (Fig. 4) but, as yet, these enzymes have not been purified and characterized. Initial cleavage is believed to occur at the N-terminal region to produce a shortened procollagen still disulphide bonded via the C-terminal extensions. Subsequent cleavage of these C-terminal peptides may involve stepwise scission of the three chains and eventually yields the fibril monomer (tropocollagen) and a trimeric disulphide bonded fragment of approx. 100 000 (25). Little is known of the fate and subsequent role, if any, of the non-helical extension peptides lost in this process.

The highly specific alignment of the fibril monomer during extracellular aggregation is critical to the formation of covalent cross-links which confer on the fibres the high tensile strength and resistance to chemical attack necessary for their function. The first step in the formation of intermolecular cross-links is the oxidative deamination of certain Lys and/or Hylys residues in the non-helical telopeptides to produce reactive aldehydes. This reaction is catalysed by lysyl oxidase (26), an extracellular copper-dependent enzyme which acts on the fibrillar substrate and may be inhibited by lathyrogens such as β-aminopropionitrile. The aldehydes produced then condense with reactive groups, predominantly the ε-$NH_2$ groups of Hylys in the helical region of neighbouring molecules to form a network of intermolecular aldimine or keto-imine bonds (for details of chemistry and biology of collagen cross-links see 27).

The above outline of the extracellular modifications occurring in

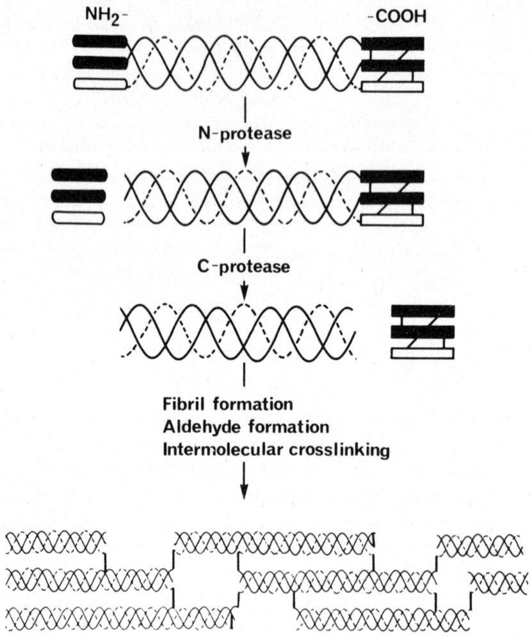


Fig. 4. Summary of events occurring during the extracellular processing of type 1 procollagen.

collagen synthesis are particularly relevant to collagen types I, II and III found in interstitial tissues. The nature and synthesis of basement membranes (BM's) containing type IV collagen is less clearly defined partly because of the difficulties in studying these specialised connective tissues. However, as noted earlier, the initially synthesised polypeptides of BM collagen have a mol. wt. of approx. 180 000 and our recent studies (10,28) on the assembly of the rat lens capsule and glomerular BM have demonstrated that these polypeptides are not subjected to extracellular conversion. If, as seems likely, the newly synthesised BM collagen has a structure similar to procollagen (Fig. 2), the retention of the extension peptides may account for the absence of distinct fibres within these matrices. Although not converted to a fibril monomer, the newly synthesised BM collagen is rapidly incorporated into material of higher mol. wt. and both the precursor polypeptides and this high mol. wt. material have been shown to be subunits of the intact lens capsule (10). The nature of this latter component is not yet certain but, since its formation can be inhibited by β-aminopropionitrile, it presumably contains Hylys-derived cross-links.

In conclusion it can be stated that considerable advances have been made in our understanding of the unique intracellular processes leading to the assembly of the procollagen molecule and the subsequent formation of collagen fibrils in the extracellular environment. However, much remains to be learned about the mechanisms controlling some of these events and particularly the regulation of collagen gene expression which is basic to normal growth and development.

Original contributions from this laboratory were supported by grants from the ARC, MRC, SRC and British Diabetic Association.

## REFERENCES

(1) Grant, M.E. & Jackson, D.S., The biosynthesis of procollagen, Essays in Biochem. 12, 77 (1976).

(2) Prockop, D.J., Berg, R.A., Kivirikko, K.I. & Uitto, J., Intracellular steps in the biosynthesis of collagen, in Biochemistry of Collagen (Ramachandran, G.N. & Reddi, A.H., eds), Plenum Press, New York, pp 163-273 (1976).

(3) Miller, E.J., Biochemical characteristics and biological significance of the genetically-distinct collagens, Molec. & Cell. Biochem. 13, 165 (1976).

(4) Kefalides, N.A., Structure and biosynthesis of basement membranes, Int. Rev. Connect. Tissue Res. 6, 63 (1973).

(5) Burgeson, R.E., El Adli, F.A., Kaitila, I.I. & Hollister, D.W., Fetal membrane collagens: Identification of two new collagen alpha chains, Proc. Nat. Acad. Sci. U.S. 73, 2579 (1976).

(6) Chung, E., Rhodes, R.K. & Miller, E.J., Isolation of three collagenous components of probable basement membrane origin from several tissues, Biochem. Biophys. Res. Commun., 71, 1167 (1976).

(7) Brown, R.A., Shuttleworth, C.A. & Weiss, J.B., Three new α-chains of collagen from a non-basement membrane source, Biochem. Biophys. Res. Commun., 80, 866 (1978).

(8) Piez, K.A., Primary Structure, In Biochemistry of Collagen Ramachandran, (G.N. & Reddi, A.H., eds), Plenum Press, New York, pp 1-44 (1976).

(9) Harwood, R., Merry, A.H., Woolley, D.E., Grant, M.E. & Jackson, D.S., The disulphide-bonded nature of procollagen and the role of the extension peptides in the assembly of the molecule, Biochem. J. 161, 405 (1977).

(10) Heathcote, J.G., Sear, C.H.J. & Grant, M.E., Studies on the assembly of the rat lens capsule. Biosynthesis and partial characterization of the collagenous component, Biochem. J., (1978) in press.

(11) Becker, U., Timpl, R., Helle, O. & Prockop, D.J., $NH_2$-Terminal extensions on skin collagen from sheep with a genetic defect in conversion of procollagen into collagen, Biochemistry 15, 2853 (1976).

(12) Olsen, B.R., Guzman, N.A., Engel, J., Condit, C. & Aase, S., Purification and characterization of a peptide from the carboxy-terminal region of chick tendon procollagen type I, Biochemistry 16, 3030 (1977).

(13) Harwood, R., Collagen polymorphism and messenger RNA, Int. Rev. Connect. Tissue Res. 8, in press.

(14) Pelham, H.R.B. & Jackson, R.J., An efficient mRNA-dependent translation system from reticulocyte lysates, Eur. J. Biochem. 67, 247 (1976).

(15) Blobel, G. & Dobberstein, B., Transfer of proteins across membranes, J. Cell Biol. 67, 835 (1975).

(16) Tryggvason, K., Risteli, J. & Kivirikko, K.I., Separation of prolyl 3-hydroxylase and 4-hydroxylase activities and the 4-hydroxyproline requirement for synthesis of 3-hydroxyproline, Biochem. Biophys. Res. Commun. 76, 275 (1977).

(17) Tuderman, L., Myllylä, R. & Kivirikko, K.I., Mechanism of the prolyl hydroxylase reaction. 1. Role of co-substrates, Europ. J. Biochem. 80, 341 (1977).

(18) Ryhänen, L., Lysyl hydroxylase. Further purification and characterization of the enzyme from chick embryos and chick embryo cartilage, Biochim. Biophys. Acta 438, 71 (1976).

(19) Ryhänen, L., Hydroxylation of lysyl residues in lysine-rich and arginine-rich histones by lysyl hydroxylase in vitro, Biochim. Biophys. Acta 397, 50 (1975).

(20) Clark, C.C. & Kefalides, N.A., Localization and partial composition of the oligosaccharide units on the propeptide extensions of type I procollagen, J. Biol. Chem. 253, 47 (1978).

(21) Kivirikko, K.I. & Myllylä, R., Collagen glycosyltransferases, Int. Rev. Connect. Tissue Res. 8, in press.

(23) Harwood, R. & Freedman, R.B., Protein disulphide isomerase activity in collagen-synthesising tissues of the chick embryo, FEBS Lett. 88, 46 (1978).

(24) Ramachandran, G.N. & Ramakrishnan, C., Molecular structure, In Biochemistry of Collagen (Ramachandran, G.N. & Reddi, A.H., eds) Plenum Press, New York, pp 45-84 (1976).

(25) Davidson, J.M., McEneany, L.S.G. & Bornstein, P., Intermediates in the conversion of procollagen to collagen, Eur. J. Biochem. 81, 349 (1977).

(26) Siegel, R.C., Lysyl oxidase, Int. Rev. Connect. Tissue Res. 8, in press.

(27) Bailey, A.J., Robins, S.P. & Balian, G., Biological significance of the intermolecular cross-links of collagen, Nature 251, 105 (1974).

(28) Heathcote, J.G., Sear, C.H.J. & Grant, M.E., Preliminary characterization of the collagenous polypeptides synthesized by lens capsules and renal glomeruli isolated from young rats, Front. Matrix Biol., (1978) in press.

# SYNTHESIS AND PROCESSING OF MILK PROTEINS

Roger Craig, Anthony Boulton and Peter Campbell
Courtauld Institute of Biochemistry,
London, U.K.

Charles Lane
N.I.M.R. Mill Hill, London, U.K.

Andrew Mellor
I.C.R.F., Lincolns Inn Fields, London, U.K.

Ponnamperuma Perera
Department of Biochemistry, University of
Sri Lanka, Sri Lanka.

Until recently the major interest in milk proteins derived from the observation that the limited number of major protein constituents present were organ specific, and that their expression was modulated by a well defined combination of both peptide and steroid hormones (for reviews see Topper & Oka (1); Banejee (2); Rosen (3)), thus providing a potentially exciting system with which to study the manner in which hormones modulate gene expression.

Experiments concerning the intracellular mechanisms involved in the synthesis and processing of secretory proteins are based fundamentally on the pioneering work of Siekevitz & Palade (4) who proposed that membrane-bound polyribosomes represented the site of synthesis of secretory proteins, whereas proteins required for intra-cellular functions (the house-keeping proteins) were synthesized on free polyribosomes. Studies designed to substantiate this concept have too often evolved around identification of the site of synthesis of immunoglobulin producing tumour-derived cell lines, which on careful comparison have produced conflicting data, both for (Ref. 5,6) and against (Ref. 7,8) this concept. However, studies using normal tissue particularly liver have gone some way to substantiate these concepts, thus there has accumulated a considerable volume of evidence which demonstrates that the major secretory component of liver, serum albumin, is synthesized primarily on membrane-bound polyribosomes, whereas ferritin, the major intracellular iron-storage protein, is synthesized primarily on free polyribosomes (see Ref. 9,10,11,12).

It has long been accepted that milk proteins are synthesized primarily on membrane-bound polyribosomes (for reviews see Denamur (13); Craig & Campbell (14)). However, in spite of the use of tissue from the normal lactating mammary gland, results concerning the precise proportion of total milk protein mRNA species associated with free or membrane-bound polyribosomes have been conflicting. Moreover as with other secretory protein systems, experiments have generally been designed to determine only the relative intracellular distribution of secretory protein or 'abundant' mRNA species and have ignored the relative distribution of the less abundant and scarce mRNA species which direct the synthesis of the many proteins not secreted by the cell. Our own preliminary observations using lactating guinea-pig mammary gland tissue, showed that after subcellular fractionation 85% of the recovered polyribosomal RNA was associated with the membrane-bound fraction, and the remainder with the free polyribosomal fraction (Ref. 15). Analysis of the ability of poly(A)-containing RNA associated with both polyribosome fractions to direct casein synthesis in a Krebs II ascites cell-free protein synthesizing system as judged by antibody precipitation, although confirming that the membrane-bound polyribosomes, were undoubtedly the predominant site of guinea-pig milk protein synthesis, also revealed considerable casein mRNA activity associated with the free polyribosome fraction. These observations were in some contrast to those of Houdebine and Gaye (16) working with the lactating ewe mammary gland, who reported the almost exclusive synthesis of casein on membrane-bound polyribosomes, as judged by the translation of casein mRNA in a reticulocyte lysate cell-free protein synthesizing system.

Since these early observations, techniques used to isolate polyribosomal fractions have improved (17), whilst the introduction of molecular hybridisation techniques, using specific cDNA probes for the quantification of specific mRNA species, has eliminated artefacts due primarily to the preferential translation of certain mRNA species, a function of the ionic conditions used for cell-free protein synthesis (Ref. 18,19). Consequently, a recent re-evaluation by Houdebine (20) of the distribution of casein mRNA between free and membrane-bound polyribosomes during lactogenesis in the rabbit, showed that 95% and 5% of total polysomal casein mRNA was found in membrane-bound and free polyribosomes respectively, whilst a considerably higher percentage of casein mRNA was associated with the free polyribosomes fraction in the pseudopregnant rabbit in the absence of prolactin.

Our own studies were designed not just to re-evaluate our previous observations, but to analyse the general distribution of polyribosome-associated mRNA species, using a combination of mRNA directed cell-free protein synthesis, polyribosome 'run-off' in the presence of an inhibitor of initiation, and comparative complexity analyses of the poly(A)-containing RNA species isolated from the free polyribosomes, membrane-bound polyribosomes and the post-nuclear supernatant of the lactating guinea-pig mammary gland. This approach (Ref. 19) demonstrated unequivocally that although as much as 40% of the poly(A)-containing RNA species associated with the free polyribosomes comprised of the abundant or milk protein mRNA species, the remainder were divided into two well defined groups, one comprising of a relatively abundant group of poly(A)-containing RNA species (450-550), and the other a large group (8000-12000) poly(A)-containing RNA species each present in only a few copies. A similar analysis of the total cytoplasmic poly(A)-containing RNA population, confirmed this general distribution, though the relative proportion of each group differed slightly, due primarily to the fact that 55% of the total population comprised of the abundant mRNA population representing the milk proteins. However, analysis of the membrane-bound poly(A)-containing RNA population revealed a considerably less complex profile. As expected the abundant milk protein mRNA species were predominant, representing over 70% of the total population, whereas the remaining poly(A)-containing RNA species appeared to comprise of a single population (2500-3500 species) of poly(A)-containing RNA molecules, a markedly less complex population than that associated with the free polyribosomes. These results supported by polyribosome 'run-off' experiments using the wheat germ cell-free protein synthesizing system *in vitro* (Ref. 19) clearly demonstrate that in the lactating guinea-pig mammary gland, there exists a very sharp functional distinction between the two polyribosomal classes. Furthermore, 'run-off' experiments conducted in the presence of aurin tricarboxylic acid (an inhibitor of initiation), suggest that significant levels of milk protein mRNA associated with the free polyribosomal preparation, may well be present as mRNP particles, and therefore causing an over estimation of the true level of milk protein mRNA species associated with the free polyribosomal preparation. However assuming that at any one time 80-85% of the total polyribosome population is associated with the endoplasmic reticulum, then in excess of 90% of the abundant milk protein mRNA population is in the form of membrane-bound polyribosomes. What do the remainder represent? It is now well established, that in addition to secretory proteins, certain integral membrane glycoproteins are also synthesized on membrane-bound polyribosomes (Ref. 22,23,24,25), thus it seems probably that the remainder of the poly(A)-containing RNA species associated with membrane-bound polyribosomes, may well direct the synthesis of either structural membrane proteins or alternatively some of the many enzymes (e.g. casein kinases) required during subsequent post-translational events involved in the secretory pathway.

Although it has been established that in the lactating mammary gland there exists a very clear functional distribution between the two polyribosome classes (Ref. 19, 20), it is of considerable importance to consider how these two types of polyribosome arise, and how those involved in secretory protein synthesis become attached to membranes, resulting ultimately in the synthesized proteins passing through the membranes into the cisternae of the endoplasmic reticulum from whence they are transported to the Golgi apparatus, packaged into secretory granules and finally released by fusion with the plasma membrane - for review see Palade (21).

As there seems little evidence to suggest that free or membrane-bound polyribosomes contain different pools of 40S and 60S ribosomal subunits (Ref. 26), it is generally accepted that some form of receptor site may exist on the endoplasmic reticulum. This will then be recognised either by the mRNA itself, or by protein(s) associated with the mRNA, or alternatively the nascent peptide. Evidence for the direct interaction of the mRNA with the membrane is conflicting (Ref. 27,28,29) though experiments performed in vivo suggest that the mRNA does not contribute significantly to the maintenance of interaction between bound polyribosomes and the membrane of the endoplasmic reticulum (Ref. 30,31). Recently, evidence has been presented (Ref. 32) which suggests that two proteins, both integral components of the rough endoplasmic reticulum appear to be related to bound polyribosomes, and therefore may be considered as possible candidates for the putative polyribosome membrane receptor site. However, evidence that the specificity of attachment, though not necessarily the mechanism of attachment resides in the nascent polypeptide is more compelling. This concept originated from observations by Milstein (33), who showed that when mRNA directing the synthesis of immunoglobin light-chain was translated in a cell-free protein synthesizing system that contained fragments of endoplasmic reticulum, and one that did not, then the product of the latter was larger, owing to the presence of an N-terminal peptide extension. These observations led to the formulation of the 'Signal' hypothesis (Ref. 33,34), which postulates the existence of a metabolically short-lived N-terminal extension to cell proteins to be segregated in membrane-bound compartments. Thus the presence of a 'Signal' pre-peptide would give rise to polyribosomes capable of attachment to the endoplasmic reticulum, via an interaction of the signal peptide with membrane receptor sites. Its absence would require complete translation of the mRNA on free polyribosomes. This hypothesis has now been elaborated (Ref. 6,35) and an accumulation of evidence has demonstrated that so far, with the major exception of the egg white protein, ovalbumin (Ref. 36), secretory proteins are synthesized with a hydrophobic signal sequence, of 18-30 amino-acid residues in length (Ref. 37,38,39,40,41, 42,43). Moreover, ingenious experiments ultilizing mRNA directing the synthesis of secretory or membrane proteins, in conjunction with cell-free protein synthesizing systems (synchronised using 7-methylguanosine 5'-phosphate) in the presence or absence of membrane fragments, have provided evidence that *in vitro*, membrane-bound ribosomes engaged in the synthesis of secretory proteins, are derived from free polyribosomes (Ref. 25,44).

It seems reasonable to predict that milk proteins should provide a suitable system with which to study intricate intracellular secretory mechanisms, particularly as mRNA species directing their synthesis have proved both easy to isolate in large quantities, and also highly active in a variety of cell-free protein synthesizing systems (see Craig & Campbell (14), Rosen (3)). Unfortunately this has not been the case. Although we demonstrated some time ago (Ref. 45), that the guinea-pig whey protein α-lactalbumin was synthesized in cell-free protein synthesizing systems lacking endoplasmic reticulum as pre-α-lactalbumin, similar positive evidence for the synthesis of pre-caseins has until recently been conspicuously lacking.

Figure 1, demonstrates the mRNA dependent synthesis of the four major guinea-pig milk proteins in the wheat germ cell-free system, in the presence of [$^{35}$S]methionine. The products have been identified using antibody precipitation, separated by SDS polyacrylamide gel electrophoresis (Ref. 46), and then visualised using fluorography (Ref. 47). The arrows denote the relative

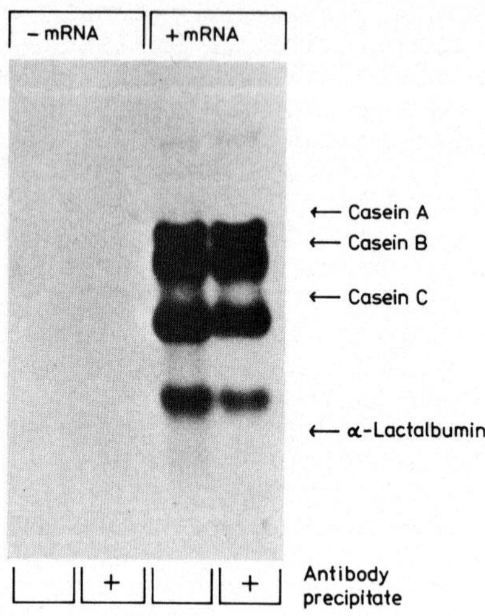


position on the gel of the marker proteins isolated from guinea-pig milk, whilst the dark bands denote the position of the in vitro synthesized antibody precipitable products. As can be clearly seen, essentially all the synthesized proteins are antibody precipitable, and none co-electrophorise with the marker proteins. One in vitro product (smallest) appears larger than the equivalent marker protein (α-lactalbumin), this in fact is pre-α-lactalbumin (Ref. 45), whilst the remainder all appear smaller than the equivalent marker caseins. Such observations appear to be common to all published data concerning the in vitro synthesis of caseins, as independent of the source of mRNA, the in vitro products all appeared to be either the same size or smaller than the authentic peptides (Ref. 45,48,49,50). However, the application of microsequencing techniques (Ref. 39) to the radiolabelled in vitro synthesized caseins has now shown that in spite of their anomalous mobility when analysed by SDS-polyacrylamide gel electrophoresis, all ovine (Ref. 51) and rat caseins (Ref. 3) examined so far, contain the 'Signal' $NH_2$-terminal peptide typical of secretory proteins.

Evidence for the presence of precursors for guinea-pig caseins, is indirect, and has been derived from comparative analysis of mRNA directed milk protein synthesis in protein synthesizing systems containing either intact endoplasmic reticulum (the oocyte) or fragments thereof. Thus analysis of milk proteins synthesized in the Krebs II ascites cell-free system which contains membrane fragments, reveals that some of the pre-α-lactalbumin has been processed into the mature protein (Ref. 45,52) whilst additional products are present representing processing of the caseins. Examination of the oocyte products reveals the presence of only the processed proteins, whereas examination of the wheat germ or reticulocyte products shows only the unprocessed proteins. Evidence that this is both a membrane function and a $NH_2$-terminal cleavage, has been obtained by comparison of the labelling pattern of the milk protein mRNA directed products, synthesized in the wheat germ cell-free system in the presence and absence of dog pancreas endoplasmic reticulum using either [$^{35}$S]Met-$tRNA_f$, [$^{35}$S]Met-tRNA or [$^{35}$S]-Methionine as the radiolabelled precursors (Ref. 52). In the presence of the latter, and added membranes, there is a marked change in the mobility of the caseins. A similar change in mobility is observed with [$^{35}$S]Met-tRNA in the presence of membranes. However, in the presence of [$^{35}$S]Met-$tRNA_f$ which will only label the initiating methionine, the addition of membrane essentially eliminates all bands from the gel, strongly suggesting that guinea-pig caseins are not only processed at the level of the endoplasmic reticulum, but that this is the result of an $NH_2$-terminal peptide cleavage. Moreover regardless of the cell-free protein synthesizing system, or the source of the added endoplasmic reticulum, whether it be from the guinea-pig mammary gland or from the dog pancreas, we have yet to obtain in any *in vitro* system caseins with the same mobility on SDS-polyacrylamide gels as the final secreted products *in vivo*. All guinea-pig caseins are phosphorylated (Ref. 56), a post-translational event known to occur in the Golgi apparatus (Ref. 53). As it seems probable that the presence of high levels of phosphate will affect the mobility of proteins on SDS polyacrylamide gels, it seems reasonable to conclude that the anomalous mobility of the *in vitro* synthesized caseins may well be due to the absence of phosphorylation *in vitro*. This conclusion is supported by our inability so far to phosphorylate (using [γ-$^{32}$P]ATP) our *in vitro* synthesized caseins, independent of the cell-free system of choice though the wheat germ in particular shows high levels of endogenous protein kinase activity.

Although we have demonstrated that milk protein processing is a membrane function, it is important to determine the intracellular fate of the products, and also to determine whether or not the passage of proteins into the cisternae of the endoplasmic reticulum is directly linked to the translational event. Experiments designed to determine the intracellular fate of secretory and non-secretory proteins synthesized by the Xenopus oocyte (Ref. 54) demonstrated that both milk protein and albumin mRNA species directed the synthesis of vesicularised products, as judged by the resistance of the newly synthesized proteins to proteinase digestion in the absence of detergent. A further series of experiments involving the microinjection of [$^{35}$S]methionine labelled milk protein precursor proteins synthesized by the wheat germ cell-free system into the oocyte, showed that less than 2% of these entered vesicles. Moreover, unlike the vesicularised proteins which are extremely stable, the unprocessed milk proteins were rapidly degraded by oocyte peptidases. In sharp contrast, the microinjection of ovalbumin synthesized in the wheat germ (this has no signal peptide) into the oocyte, resulted in minimal degradation of the protein (Ref. 55).

These observations are interesting in that (i) they demonstrate in a essentially in vivo situation, the absolute requirement for active ongoing translation before selective compartmentalisation of secretory proteins may occur, even in the presence of the signal peptide and (ii) raise the possibility for an alternative or additional role for the signal peptide, in that should secretory proteins bearing signal sequences arise in the wrong compartment of the cell, then these are rapidly degraded, effectively preventing the accumulation of the primary translation products.

The results of our investigations and those of others into the intracellular mechanisms involved in milk-protein secretion, confirm that a very clear functional distinction exists between the two classes of polyribosomes. Precisely how these arise has still to be established. Our evidence is consistent with the initiation of all protein synthesis on free polyribosomes, followed by the attachment of a discrete population of polyribosomes to the endoplasmic reticulum. It seems reasonable to predict that the nascent 'Signal' sequence plays a role in the specificity of these events, involving (i) the initial interaction between the 'free' polyribosome and the putative membrane receptor sites, and (ii) possibly at a different level preventing the miscompartmentalisation of those secretory proteins bearing such a sequence.

We thank the Medical Research Council for supporting this work. Dr. P.A.J. Perera was supported by a Commonwealth Medical Fellowship.

REFERENCES

(1) Y.J. Topper and T. Oka, Some aspects of mammary gland development in the mature mouse, Lactation 1, 327 (1974).

(2) M.R. Banerjee, Responses of mammary cells to hormones, Int. Rev. Cytol. 47, 1 (1976).

(3) J.M. Rosen, Gene expression in normal and neoplastic breast tissue, Breast Cancer 2. W.L. McGuire, ed. Plenum Press, New York (in press).

(4) P. Siekevitz and G.E. Palade, A cytochemical study on the pancreas of the guinea-pig, J. Biophys. Biochem. Cyt. 7, 619 (1960).

(5) D. Cioli and E.S. Lennox, Immunoglobulin nascent chains on membrane-bound ribosomes of myeloma cells, Biochemistry 12, 3211 (1973).

(6) G. Blobel and B. Dobberstein, Transfer of proteins across membranes. I. J. cell Biol. 67, 835 (1975).

(7) B. Lisowska-Bernstein, M.E. Lamm and P. Vassalli, Synthesis of immunoglobulin heavy and light chains by the free ribosomes of a mouse plasma cell tumor, Proc. Natl. Acad. Sci. 66, 425 (1970).

(8) A. Okuyama, J. McInnes, M. Green and S. Pestka, Distribution of MOPC-315 light chain messenger RNA in free and membrane-bound polyribosomes, Biochem. Biophys. Res. Commun. 77, 347 (1977).

(9) S.J. Hicks, J.W. Drysdale and H.N. Munro, Preferential synthesis of ferritin and albumin by different population of liver polysomes, Science 164, 584 (1969).

(10) J. Zahringer, B.S. Baliga, R.L. Drake and H.N. Munro, Distribution of ferritin mRNA and albumin mRNA between free and membrane-bound rat liver polysomes, Biochim. Biophys. Acta 474, 234 (1977).

(11) C.M. Redman, The synthesis of serum proteins on attached rather than free ribosomes of rat liver, Biochem. Biophys. Res. Communs. 31, 845 (1968).

(12) S.H. Yap, R.K. Strair, and D.A. Shafnitz, Distribution of rat liver albumin mRNA membrane-bound and free in polyribosomes as determined by molecular hybridisation, Proc. Natl. Acad. Sci. 74, 5397 (1977).

(13) R. Denamur, Ribonucleic acids and ribonucleoprotein particles of the mammary gland, Lactation 1, 413 (1974).

(14) R.K. Craig and P.N. Campbell, Molecular aspects of milk protein biosynthesis, Lactation 4, 387 (1978).

(15) O.S. Harrison, R.K. Craig and P.N. Campbell, Isolation and characterization of messenger ribonucleic acid species for guinea-pig milk proteins from free and membrane-bound polyribosomes, Biochem. Soc. Trans. 4, 340 (1976).

(16) L.M. Houdebine and P. Gaye, Absence of mRNA for casein in free polysomes of lactating ewe mammary gland, Nucleic Acid Res. 2, 165 (1975).

(17) J.C. Ramsey and W.J. Steele, Differences in size, structure and function of free and membrane-bound polyribosomes of rat liver, Biochem. J. 168, 1 (1977).

(18) D.A. Shafritz, Molecular mechanisms of protein biosynthesis, in Molecular Biology Series (1977) Academic Press, New York.

(19) R.K. Craig, O.S. Harrison, A.P. Boulton and P.N. Campbell, Manuscript in Preparation (1978).

(20) L.M. Houdebine, Distribution of casein mRNA between free and membrane-bound polysomes during the induction of lactogenesis in the rabbit, Mol. Cell. Endocrinol., 7, 125 (1977).

(21) G. Palade, Intracellular aspects of the process of protein synthesis, Science 189, 347 (1975)

(22) M.J. Grubman, E. Ehrenfeld and D.F. Summers, In vitro synthesis of proteins by membrane-bound polyribosomes from vesicular stomatitis virus-infected Hela cells, J. Virol. 14, 560 (1974)

(23) T. Morrison and H.F. Lodish, Site of synthesis of membrane and nonmembrane proteins of vesicular stomatitis virus, J. Biol. Chem. 250, 6955 (1975).

(24) D.F. Wirth, F. Katz, B. Small and H.F. Lodish, How a single Sindbis virus mRNA directs the synthesis of one soluble protein and two integral membrane glycoproteins, Cell, 10, 253 (1977).

(25) J.E. Rothman and H.F. Lodish, Synchronised transmembrane insertion and glycosylation of a nascent membrane protein, _Nature_, 269, 775 (1977).

(26) J.A. Lewis and D.D. Sabatini, Proteins of rat liver free and membrane-bound ribosomes. Modification of two large subunit proteins by a factor detached from ribosomes at high ionic strength, _Biochim. biophys. Acta_ 478, 331 (1977).

(27) M.A. Lande, M. Adesnik, M. Sumida, Y. Tashiro and D.D. Sabatini, Direct association of messenger RNA with microsomal membranes in human fibroblasts, _J. cell Biol_. 65, 513 (1975).

(28) M. Adesnik, M.A. Lande, T. Martin, and D.D. Sabatini, Retention of mRNA on the endoplasmic reticulum membranes after _in vivo_ disassembly of polysomes by an inhibitor of initiation, _J. cell Biol_. 71, 307 (1976).

(29) J. Cordelli, B. Long and H.C. Pitot, Direct association of messenger RNA labelled in the presence of fluoroorotate with membranes of the endoplasmic reticulum in rat liver, _J. cell. Biol_. 70, 47 (1976).

(30) H.F. Lodish and S. Froshauer, Binding of viral glycoprotein mRNA to endoplasmic reticulum membranes disrupted by puromycin, _J. Cell Biol_. 74, 358 (1977).

(31) J. Kruppa and D.D. Sabatini, Release of poly A (+) messenger RNA from rat liver rough microsomes upon disassembly of bound polysomes, _J. cell. Biol_. 74, 414 (1977).

(32) G. Kreibich, C.M. Freienstein, B.N. Pereyra, B.L. Ulrich and D.D. Sabatini, Proteins of rough microsomal membranes related to ribosome binding. _J. cell Biol_. 78, 488 (1978).

(33) C. Milstein, G.G. Brownlee, I.M. Harrison and M.B. Mathews, A possible precursor of immunoglobulin light chains, _Nature New Biol_. 239, 117 (1972).

(34) G. Blobel and D.D. Sabatini, Ribosome membrane interaction in eukaryotic cells, _Biomembranes_ 2, 193 (1971).

(35) G. Blobel and B. Dobberstein, Transfer of proteins across membranes II. _J. cell Biol_. 67, 852 (1975).

(36) R.D. Palmiter, J. Gagnon and K.A. Walsh, Ovalbumin: A secreted protein without a transient hydrophobic leader sequence, Proc. Natl. Acad. Sci. 75, 94 (1978).

(37) A. Devillers-Thiery, T. Kindt, G. Scheele and G. Blobel, Homology in amino-terminal sequence or precursors to pancreatic secretory proteins, Proc. Natl. Acad. Sci. 72, 5016 (1975).

(38) Y. Burstein and I. Schechter, Amino acid-sequence variability at the N-terminal extra piece of mouse immunoglobulin light-chain precursors of the same and different subgroups, Biochem. J. 157, 145 (1976).

(39) N. Kemper, J.F. Habener, M.D. Ernst, J.T. Potts, Jr., and A.Rich, Pre-proparathyroid hormone: Analysis of radioactive tryptic peptides and amino acid sequence, Biochemistry 15, 5 (1976).

(40) V.R. Lingappa, A. Devilliers-Thiery and G. Blobel, Nascent prehormones are intermediates in the biosynthesis of authentic bovine pituitary growth hormone and prolactin, Proc. Natl. Acad. Sci. 74, 2432 (1977).

(41) R.A. Maurer, J. Gorski and D.J. McKean, Partial amino acid sequence of rat pre-prolactin, Biochem. J. 161, 189 (1977).

(42) A.W. Strauss, A.M. Donohue, C.D. Bennett, J.A. Rodbey and A.W. Alberts, Rat liver preproalbumin: *In vitro* synthesis and partial amino acid sequence, Proc. Natl. Acad. Sci. 74, 1358 (1977).

(43) S.J. Chan, P. Keim, and D.F. Steiner, Cell-free synthesis of rat preproinsulins: Characterization and partial amino acid sequence determination, Proc. Natl. Acad. Sci. 73, 1964 (1976).

(44) I. Boime, E. Szczesna and D. Smith, Membrane dependent cleavage of the human placental lactogen precursor to its native form in ascites cell-free extracts, Eur. J. Biochem. 73, 515 (1977).

(45) R.K. Craig, P.A. Brown, O.S. Harrison, D. McIlreavy and P.N. Campbell, Guinea-pig milk protein synthesis. Isolation and characterization of messenger ribonucelic acids from lactating mammary gland and identification of caseins and pre-$\alpha$-lactalbumin as translation products in heterologous cell-free systems, Biochem. J. 160, 57 (1976).

(46) K. Weber, J.R. Pringle and M. Osborn, Measurement of molecular weights by electrophoresis on SDS-acrylamide gel, Methods Enzymol. 26C, 3 (1972).

(47) W.M. Bonner and R.A. Laskey, A film detection method for tritium-labelled proteins and nucleic acids in polyacrylamide gels, Eur. J. Biochem. 46, 83 (1974).

(48) P. Gaye and L.M. Houdebine, Isolation and characterization of casein mRNA's from lactating ewe mammary glands, Nucleic Acid Res. 2, 707 (1975).

(49) J.M. Rosen, S.L.C. Woo and J.P. Comstock, Regulation of casein messenger RNA during the development of the rat mammary gland, Biochemistry, 14, 2895 (1975).

(50) J.M. Rosen, Isolation and characterization of purified rat casein messenger ribonucleic acids, Biochemistry 15, 5263 (1976).

(51) P. Gaye, J.P. Gautron, J-C. Mercier, and G. Haze., Amino terminal sequences of the precursors of ovine caseins, Biochem. Biophys. Res. Comm. 79, 903 (1977).

(52) R.K. Craig, A.P. Boulton, P.A.J. Perera, A. Mellor and C. Lane - Manuscript in Preparation (1978).

(53) A.G. Mackinlay, D.W. West and W. Manson, Specific casein phosphorylation by a casein kinase from lactating bovine mammary gland, Eur. J. Biochem. 76, 233 (1977).

(54) T. Zehavi-Willner and C. Lane, Subcellular compartmentation of albumin and globin made in oocytes under the direction of injected messenger RNA, Cell 11, 683 (1977).

(55) C. Lane and R.K. Craig, Manuscript in preparation (1978).

(56) R.K. Craig, D. McIlreavy and R. Hall, Separation and partial characterization of guinea-pig caseins, Biochem. J. 173, 633 (1978).

# SYNTHESIS AND PROCESSING OF IMMUNOGLOBULINS

Fritz Melchers
Basel Institute for Immunology, Basel, Switzerland

## ABSTRACT

Immunoglobulins (Ig) are produced by B-lymphocytes. B-lymphocytes develop from stem cells in distinct stages, some of which can be characterized by their mode of immunoglobulin synthesis, turnover, surface membrane deposition and secretion. The earliest Ig-synthesizing cell, developing at day 11 of embryonic development of the mouse, synthesizes 7-8 S IgM which it rapidly turns over. Turnover is by release from the cell, a process which at some stage locates Ig in the surface membrane. From these earliest Ig-producing, large-size B-cells, small resting B-cells develop which also synthesize 7-8 S IgM but turn over this IgM more slowly. Turnover is by shedding from the cell surface. Binding of antigen or mitogen to surface Ig and mitogen-receptors in the surface membrane initiates growth and differentiation of B-cells to IgM-, IgA- and IgG-secreting cells, so-called plasma cells, which now actively secrete 19 S IgM, monomeric and polymeric IgA and 7 S IgG from the cells. Turnover of the secreted Ig is rapid; plasma cells cease to insert Ig in their surface membrane. Active secretion involves the synthesis of heavy and light chains of Ig on the intracytoplasmic side of the rough endoplasmic reticulum (R-ER) on membrane-bound polyribosomes, the transfer of these chains through the lipid bilayer of the membranes of the R-ER to the other, extracytoplasmic side, the migration of the Ig from the R-ER to the smooth ER and from there out of the cell. Concomitant with this migration a stepwise attachment of carbohydrate to heavy chains of Ig occurs at different subcellular sites. Synthesis and processing of Ig changes during differentiation of B-cells and can, therefore, be taken as a marker for B-cell differentiation.

F. Melchers

## INTRODUCTION

Immunoglobulins (Ig) are synthesized by B-lymphocytes (1). All experimental evidence supports the view that no other cell in the body synthesizes Ig and that Ig-synthesis is, therefore, a special, differentiated function of B-cells. Expression of Ig synthesis appears to occur only after separate genes for variable (V-) and constant (C-) regions of Ig-heavy (H-) and light (L-) chains have been joined on the DNA level (2). This joining of V- and C-genes is believed to occur within the development of B-lymphocytes from pluripotent stem cells, first during embryonic development, and later continuously during ontogeny of B-cells in the bone marrow of adult animals (1, 3-6). The function of Ig molecules, more precisely of V-regions of H- and L-chains, is to bind determinants of antigens. Generally Ig molecules can occur in two forms to bind antigens: as soluble molecules in serum or as surface-membrane-bound constituents of cells. Before B-cells can recognize antigen they have to insert the synthesized Ig molecule in the surface membrane. This recognition of antigen leads to proliferation of B-cells, and to maturation of such stimulated cell clones into Ig-secreting cells. The phenotypic expression of Ig-synthesis, therefore, changes after stimulation from a mode of synthesis in which Ig is inserted in the surface membrane, to another mode, in which Ig molecules are secreted from the cells into the bloodstream or, in vitro, into the supernatant medium.

Binding of antigen to surface-membrane bound Ig molecules on B-cells is, however, insufficient to trigger the reactions leading to growth and to secretion. Second types of receptors, so-called mitogen-receptors, have to be occupied by growth-inducing mitogens in order to initiate growth and secretion. Mitogens for B-cells can either be external in nature - such as lipopolysaccharide (LPS) or lipoprotein (LPP) from the surface membrane of gram-negative bacteria (7,8) - or can be supplied within the immune system by cooperating T-cells (1). A complex of Ig-molecules and mitogen receptors, in which the occupance by either antigen or anti-Ig-molecules can modulate the reactions triggered by the occupance of mitogen-receptors with mitogen, is implied by functional studies (9). This association must be weakened and finally lost as mitogen- and antigen-reactive B-cells mature to end-stage, Ig-secreting plasma cells. As maturation develops in B-cells with time of stimulation the representation of Ig on the surface of the cells is lost. This leads one to speculate that Ig-molecules to be secreted from plasma blasts amd mature plasma cells associate themselves with other components of the cells which then constitute the secreting pathway.

Insertion of Ig into the surface membrane and secretion of Ig from the activated, matured cells, are both processes in which Ig is believed to be membrane-bound. It should, therefore, be the neighbouring molecules-proteins, carbohydrates or lipids - which determine the way in which Ig will proceed after synthesis. Our understanding of the process of surface membrane insertion and of active secretion would probably increase if we knew the genes, the molecular nature and the regulation of phenotypic expression of these neighbouring molecules. The biochemical search for mitogen receptors on B-cells may be a first step in this direction.

## IMMUNOGLOBULINS SYNTHESIZED IN DIFFERENT B-CELL SUBPOPULATIONS

Analyses of the biochemical modes of synthesis and processing of Ig are complicated by the fact that every normal lymphoid organ of the immune system consists of a mixture of different T- and B-cell populations at different stages of development. The relative content of all such different subpopulations of lymphocytes even changes with the age of the animal. Separations according to physicochemical properties of B-cells, i.e., size and surface charge, according to functional properties, i.e., reactivities to mitogens, to T-cell independent and to T-cell dependent antigens, and according to their time of first appearance during embryonic development have led us to distinguish five different types of B-cells (10,11). Their properties are given in Table 1.

These five different types of B-cells are certainly only a minimal number. In fact, further distinctions can be made within the small, resting B-cells (type 3, Table 1) and the plasma blasts and plasma cells (types 4 + 5, Table 1). Small, resting B-cells can express more than one isotype of Ig, i.e. more than one class of H-chain C-regions per cell. There exist surface Ig-positive, antigen- and mitogen-reactive small, resting B-cells which express on their surface either a) only μ-H-chains, b) μ- and γ-chains, c) μ and δ chains, d) μ-, γ- and δ-chains, e) only γ-chains or f) only δ-chains. Further differences are found when the subclasses of γ-H-chains (γ1, γ2, γ3, γ4) and when α-chains are included in these distinctions. During B-cell development μ-H-chains appear first, followed by the expression of γ- and δ-chains (12-14). As a rule these different H-chain C-regions expressed on one B-cell carry identical $V_H$-regions. At the level of plasmablast and plasma cell development (types 4 and 5, Table 1) differences in isotype, H-chain class expression can, again, be found. Ninety percent of all resting B-cells of a 3 months-old mouse will only develop into IgM-secreting plasma cells after mitogenic stimulation, while 10% of the mitogen-reactive B-cells switch the expression of the H-chain class to γ-chains (16). Again, the rule appears to hold that one clone of switching cell expresses the same $V_H$-region on two different $C_H$-regions (Cμ and Cγ). Mitogenic reactivities are other parameters by which different small, resting B-cell subpopulations can be further distinguished. At birth small B-cells of the mouse, in fetal liver, are reactive only to LPS but not to LPP. At one year of age, B-cells in spleen are no longer reactive to either LPS or LPP, but only to T-cell help. At three months of age most small B-cells are reactive to more than one B-cell mitogen -LPS and LPP as well as to T-cell help (11,15). These findings imply that the repertoire of mitogen-receptors is different on different B-cell subpopulations and that it changes with age.

TABLE 1 Properties of B-cell Subpopulations in the Mouse

| Type | Common name | Predominant occurrence | Biochemical characteristics of Ig-synthesis |
|---|---|---|---|
| 1 | large pre B-cell | fetal liver (day 11 to 15 of gestation), bone marrow | Synthesis of 7-8 S IgM with rapid turnover (t 1/2∿45 min). Intermediate surface representation of IgM. Not reactive to external mitogens or to external antigens. |
| 2 | small pre B-cell | fetal liver (day 16 to 19), bone marrow | Synthesis of IgM of 7-8 S IgM with slow turnover (t 1/2∿20 hr). Surface deposition of IgM. Cells <u>not</u> reactive to external mitogens. |
| 3 | small resting B-cell | 90% of normal B-cells in spleen, lymph nodes or thoracic duct. | Synthesis of 7-8 S IgM will slow turnover (t 1/2∿20 hr). Surface deposition of IgM. Cells reactive to external mitogens and antigens. |
| 4 | plasmablast | after stimulation <u>in vivo</u> or <u>in vitro</u> of resting B-cells within the first 48 hr of stimulation | Synthesis of 7-8 S IgM with slow turnover (t 1/2∿20 hr) and of 19 S IgM with rapid turnover (t 1/2∿2-4 hr). Surface deposition <u>and</u> secretion of IgM. Cell reactive to external mitogens and antigens. |
| 5 | plasma cell | after stimulation <u>in vivo</u> or <u>in vitro</u> of resting B-cells at later times (4-8 days) | Synthesis of 19 S IgM with rapid turnover (t 1/2∿2-4 hr). Active secretion of 19 S IgM. Cells no longer reactive to external mitogens and antigens. Switch to IgG secretion. |

## MALIGNANT STATES OF B-CELL SUBPOPULATIONS

The bewildering heterogeneity of B-lymphocytes, quite apart from their even more bewildering capacity to synthesize the many different ($>10^7$) $V_H/V_L$-combinations of antigen-binding sites on Ig molecules, can be resolved, wherever clones of B-cells can be expanded to large numbers of identical cells. Normal B-cells are not suited to such large scale clonal expansions for a number of reasons: a) they are either resting cells which must be stimulated to grow , b) when they grow they often differentiate into new types of cells (17), c) often, the growth-inducing stimulator (mitogen) is unknown. This is particularly true for type 1 and 2 pre B-cells (Table 1) and d) the end-stage plasma cell looses the capacity to grow.

For biochemical investigations, malignant forms of B-cells have long been the best and only source for clones of B-cells. Plasmacytomas producing different classes of Ig-molecules (18) appear arrested at different stages of their differentiation from resting cells (19), with an apparently unlimited capacity to divide. They facilitate the comparative biochemical analysis with the normal plasma cell and with the normal resting cell and, thereby, may illuminate the molecular and cellular processes which transform a normal cell into a malignant state.

While plasma cells and plasmablasts have plasmacytomas as malignant counterparts we are less certain of defining properties of B-cell malignancies, such as chronic lymphocytic leukemias and Abelson-virus-transformed (in the mouse) or Epstein-Barr-virus-transformed (in man) B-cell lines which would match those of either type 1, 2 or 3 normal B-cells. This is mainly so, because the normal cells remain resting, while the above mentioned malignancies divide continuously.

## Ig-SYNTHESIS AND PROCESSING IN PLASMA CELLS

Ig is synthesized on membrane-bound polyribosomes which sit on that side of the rough endoplasmic reticulum membrane that faces the intracellular compartment around the nucleus of the cell. New synthesized chains include pre-sequences at the N-terminal ends, which are composed mainly of hydrophobic amino acids (20-23). These hydrophobic amino acids are thought to guide the immunoglobulin chains through the hydrophobic lipid bilayer of the RER to the other side, i.e., into the channels of the RER (24,25). The pre-pieces are split off at this point and the Ig molecules continue to migrate from the RER to the SER before being secreted. In an IgM-secreting plasma cell, between 5 and 40% of all protein can be identified as IgM, while outside the cell IgM constitutes 85-100% of the secreted protein. Plasma cells, therefore, appear specialized to secrete Ig.

Different classes of Ig are synthesized and transported in plasma cells in a similar fashion. It appears that they all move from their site of synthesis on membrane-bound polyribosomes into the cisternae of the RER. The migration of Ig from the RER to the SER varies in different types of plasma cells. IgM in many plasmablasts has a very small pool in the SER. Therefore, it moves very fast from the RER to the outside of the cell. More mature IgM-secreting plasma cells (19) have a longer period of migration through the SER, thus IgM can be detected in the SER in such cells. Most IgG and IgA-secreting plasma cells have an easily detactable larger pool in the SER. SER is, therefore, a detectable site of migration of IgG and IgA from RER to the outside of the cell (26-30). It appears that the development of a pool of Ig-molecules in the SER parallels an increased maturation from plasmablastoid cells to mature plasma cells, morphologically detectable by the development and extension of endoplasmic reticulum membranes. The H- and L-chains of different classes of Ig are assembled at different subcellular sites. This is particularly evident when the assembly of the disulphide bonds is monitored (31). Up to the end of their migration through the RER, IgM-molecules are only assembled to 7-8 S subunits. In most plasmablasts IgM is then rapidly secreted. Shortly before or during secretion it is assembled to 19 S pentamers (32). J-chains are added to the pentamers at that time. It appears that in more mature IgM-secreting cells, 19 S IgM molecules can be found inside the cell. So far, it has not been possible to trace their location to a specific subcellular site, although the SER appears most likely to initiate this intracellular pentamerization.

Ig molecules contain carbohydrate residues, most of them attached in an N-glycosidic linkage through asparagine-N-acetylglycosamine. IgM contains five such carbohydrate groups attached per H-chain, most IgG and IgA molecules have only one per H-chain. L-chains are usually devoid of carbohydrate. Radiochemical analyses of IgM-molecules labelled by radioactive precursors of carbohydrate moieties have permitted the qualitative analysis of the presence or absence of certain sugar residues in Ig (26,27,34). Analysis of the composition and sequence of sugars in these carbohydrate groups have clarified their structure (33). Mannose and glucosamine residues constitute the <u>core</u> of the carbohydrate groups. They are added at an early stage in the process of transport and secretion, after or during the release from the membrane-bound polyribosomes within the RER. Mannose-N-acetyl-glucosamine-oligosaccharide-dolichol phosphates act as the first intermediates in the glycosylation reaction (35). The core of mannoses and N-acetylglucosamines, therefore, appears to be presynthesized and then added as one big group to Ig. Galactose, fucose and N-glycolyl-neuraminic acid residues are penultimate and terminal sugar residues of the carbohydrate groups. They are added, apparently as single units, while the Ig-molecules are traversing the SER (galactoses, N-glycolyl-neuaminic acids) or just before secretion (fucoses). Thus during migration through the membraneous structures of a plasma cell Ig acquires, by stepwise addition, the residues of its carbohydrate moieties.

These findings have suggested the hypothesis that the biosynthetic steps by which carbohydrate residues are attached to Ig might be part of the process by which plasma cells transport and secrete Ig

(36,37). Glycosyltransferases, located in membranes of the rough and smooth endoplasmic reticulum, could attach sugar residues to Ig in an ordered fashion and thereby constitute an assembly line which directs Ig molecules out of the cells. This possible biological role for carbohydrate moieties of Ig in selection for transport and secretion could be tested with an inhibitor of carbohydrate attachment to glycoproteins, e.g. 2-deoxy-D-glucose (38-40). IgG-producing tumour cells and IgM-producing tumour cells and mitogen-activated cells were tested (29,30). 2-Deoxy-D-glucose inhibited attachment of carbohydrate to Ig in tumour plasma cells and in mitogen-stimulated B-cells. In both types of cell 2-deoxy-D-glucose did not inhibit synthesis of μ-chains, γ-chains and L-chains, but inhibited active secretion of these newly made molecules. On polyacrylamide-gel electrophoresis, mobility of the μ-chains synthesized in the presence of the inhibitor was that expected for a μ-chain with less or no carbohydrate attached to them.

Differences were observed in the active secretion of IgM molecules from deoxy-D-glucose-inhibited cells, which had been synthesized before the addition of the inhibitor. In tumour plasma cells approx. 30% of the intracellular IgM can be secreted as 19 S pentamers in the presence of the inhibitor. This indicates that sugar residues, such as galactose and fucose, normally added to these molecules shortly before secretion from the cells are not requisite for the pentamerization of IgM molecules. This makes it unlikely that addition of these carbohydrate residues is primarily necessary for induction or stabilization of any conformational changes between 7-8 S intracellular and 19 S extracellular IgM. In contrast with tumour plasma cells, mitogen-stimulated B-cells did not secrete any IgM as 19 S pentamers in the presence of the inhibitor (30). Instead IgM molecules, which normally disappear from the cells at a rapid rate to be polymerized into 19 S pentamers, now appeared aggregated into molecules much larger than 19 S. It is likely that mitogen, in the presence of the inhibitor, induces this aggregation of IgM subunits on the surface of the stimulated 2-deoxy-D-glucose-inhibited B-cells, in a way analogous to the observed aggregation of surface IgM in B-cells in the initial phase of mitogen stimulation. We take these findings as additional evidence that IgM may stay temporarily on the surface membrane during active secretion.

In IgG-producing and -secreting tumour plasma cells, 2-deoxy-D-glucose inhibited the migration of newly synthesized IgG polypeptide chains from membrane-bound polyribosomes into the cisternae of the RER. It also inhibited the transfer of IgG molecules from the RER to the SER. It does not inhibit the transfer of IgG molecules from the SER to the outside of the cells. 2-Deoxy-D-glucose inhibited the attachment of sugar residues to IgG. Thus attachment of galactose, fucose and N-glycolylneuraminic acid to IgG molecules located in the SER is not a prerequisite for their secretion from plasma cells. Glycosylation of IgG molecules and/or other intracellular carbohydrate moieties is, however, necessary to draw newly synthesized IgG molecules into rough membranes and to transport them from there into smooth membranes (29). More recently the antibiotic tunicamycin has been employed to investigate the role of glycosylation in the secretion of Ig from plasma cells (41,42). Tunicamycin selectively prevents the glycosylation of newly-synthesized proteins by inhibiting the formation of N-acetyl-glucosamine-lipid-intermediates (43). It was found that secretion of IgE was most strongly inhibited (over 90%), followed by inhibition of IgM secretion (80%), IgA-secretion (65%) and IgG or λ-L-chain secretion (30%) (44,45). Subcellular migration has not yet been studied extensively in these conditions. So far, it appears that 2-deoxy-glucose and tunicamycin have very similar effects in inhibiting Ig-secretion.

Secretion of Ig light-chain subunits without detectable carbohydrate attached to them (46) and secretion of full IgG molecules with incomplete carbohydrate moieties (27), could already be taken as evidence against the hypothesis that glycosylation of Ig molecules is requisite for Ig secretion. We have found (F. Melchers and J. Andersson, unpublished work) that active secretion of the carbohydrate-free light chain of the plasma cell tumour MOPC 41 is inhibited by 2-deoxy-D-glucose. Since these tumour cells do not produce any detectable carbohydrate-containing Ig chain, such as heavy chain, we conclude from these experiments that glycosylation of intracellular structures other than Ig are requisite to draw newly synthesized Ig molecules into membranous channels of the RER and from there into those of the SER. More evidence has recently been accumulated that glycosylation of Ig (and of other secreted glycoproteins) is not directly, but may be indirectly involved in processes of active secretion. Mutants of myeloma cells have been isolated in which 30-50% of all synthesized H-chains are, nonetheless, assembled, transported and secreted with similar kinetics to the glycosylated H-chains in the same cells (47). It suggests that, if glycosylation is important for secretion, non-glycosylated H-chains could be carried along with glycosylated H-chains, and that secretion of Ig is a process which could involve many molecules in one complex.

## Ig-SYNTHESIS AND PROCESSING IN SMALL, RESTING B-CELLS

The mode of synthesis, transport and surface deposition of Ig has been studied in resting, small B-cells from spleen, lymph node, thoracic duct, bone marrow and fetal liver. Although it is almost certain that such small cells represent a mixture of different B-cells (see above and Table 1) the mode of IgM synthesis, turnover, carbohydrate attachment and surface deposition appears at least very similar in all of the different subpopulations of small B-cells.

A single resting, small B-cell contains between 3 and 15 x $10^4$ Ig molecules. Over 85% of them are located in the surface membrane.

These Ig molecules turn over with an approximate half-time (t 1/2) of 20 hr. On the surface they are 7-8 S IgM subunits. Turnover is by a process called "shedding" in which 7-8 S IgM molecules, most probably together with other membrane molecules are released essentially undegraded into the supernatant medium. 7-8 S IgM in the surface membrane contains only core, but not branch sugars (galactoses, fucoses, N-acetyl-neuraminic acids). No apparent differences in the size of the polypeptide portions of Hμ- and L-chains have been found between surface-bound and shed 7-8 S IgM from small cells, and intracellular 7-8 S IgM and secreted 19 S IgM from plasma cells (48). Active secretion of 19 S IgM is absent in small B-cells.

Stimulation of small, resting B-cells by mitogens or antigens change the mode of Ig-synthesis very rapidly. Surface-bound IgM aggregates, and is thereafter degraded by apparently surface membrane-bound proteases (30). The rate of IgM synthesis increases within the first 4 hr of stimulation. This IgM synthesized at an increased rate shows high turnover (t 1/2∿4 hr), is pentamerized to 19 S IgM outside the cells, contains branch sugars and, therefore, shows all characteristics of actively secreted IgM (49,50). Reprogramming of preexisting messenger RNA for Hμ-chain-synthesis from a synthesis of membrane-bound to a synthesis of secreted IgM appears to occur. As time of stimulation continues the ratios of the rates of synthesis and secretion of IgM over those of all proteins made in the cell continue to increase. Branch sugars continue to be added to 19 S IgM molecules actively secreted from the activated plamablasts and plasma cells. In fact synchronization of activated cells (W. Lernhardt and F. Melchers, in preparation) have revealed that 19 S IgM with branch sugars is synthesized and secreted during successive cell cycles at ever increasing rates in the Gl phase of the cycle. A balance appears to exist in stimulated cells between reactions leading to growth and those leading to differentiation. Those characterized for differentiation, i.e., branch sugar addition to 19 S IgM with rapid turnover, increase in rate with succeeding cell cycles and, thereby, imbalance the cells toward more differentiation until they cease to grow.

## Ig-SYNTHESIS AND PROCESSING IN PRECURSOR B-CELLS

The earliest cells synthesizing Ig in the mouse can be detected by biosynthetic incorporation of leucine and by lactoperoxidase-catalyzed radioiodination of fetal liver cells, followed by serological precipitation and identification as 7-8 S IgM by gel electrophoresis from day 10 to day 12 of gestation onward. These early Ig-synthesizing cells are large; they display their Ig on the surface but release it rapidly (t 1/2 ∿ 45 min). These cells may be identical with pre-B cells identified by adoptive transfer experiments to be precursors of antigen-reactive B cells. Up to day 15 of gestation, such large precursor B cells are only detectable Ig-synthesizing B cells in fetal liver. Thereafter, small, surface-Ig-positive cells appear which release 7-8 S IgM slowly. Fetal liver cells, put in culture at day 14 of gestation, will change their rate of IgM synthesis and turnover from rapid (t 1/2 = 45 min) to slow (t 1/2 = 20 hr) within 3 days in vitro, indicating that such large pre-B cells may differentiate to small B cells in culture (11). Nothing is known of the mode of transport or processing of Ig-molecules in these early B-cells.

The limited information on the molecular and kinetic parameters of Ig synthesis and turnover, however, make it already feasible to delineate a pathway of differentiation of B-cells from pre B-cells (type 1, Table 1) to plasma cells (type 5, Table 2) which can be characterized by the size and turnover of IgM molecules synthesized at the different stages of B-cell development.

## REFERENCES


(1) Miller, J.F.A.P. & Mitchell, G.F. Thymus and antigen-reactive cells. Transplant. Rev. 1, 3 (1969).

(2) Tonegawa, S., Hozumi, N., Matthyssens, G. & Schuller, R. Somatic Changes in the content and context of immunoglobulin genes. Cold Spring Harbor Symp. Quant. Biol. 41, 877 (1977).

(3) Tyan, M.L. & Herzenberg, L.A. Studies on the ontogeny of the mouse immune system. II. Immunoglobulin-producing cells. J. Immunol. 101, 446 (1968).

(4) Tyan, M.L., Ness, D.B. & Gibbs, P.R. Fetal lymphoid tissues: antibody production in vitro. J. Immunol. 110, 1170 (1973).

(5) Brahim, F. & Osmond, D.G. The migration of lymphocytes from bone marrow to popliteal lymph nodes demonstrated by selective bone marrow labelling with [$^3$H]thymidine in vivo. Anat. Rec. 175, 737 (1973).

(6) Osmond, D.G. & Nossal, G.J.V. Differentiation of lymphocytes in mouse bone marrow. II. Kinetics of maturation and renewal of antiglobulin-binding cells studied by double labelling. Cell. Immunol. 13, 132 (1974).

(7) Andersson, J., Sjöberg, O. & Möller, G. Induction of immunoglobulin and antibody synthesis in vitro by lipopolysaccharides. Europ. J. Immunol. 2, 349 (1972a).

(8) Melchers, F., Braun, V. & Galanos, C. The lipoprotein of the outer membrane of Escherichia coli: A B-lymphocyte mitogen. J. exp. Med. 142, 473 (1975).

(9) Andersson, J., Bullock, W.W. & Melchers, F. Inhibition of mitogenic stimulation of mouse lymphocytes by anti mouse immunoglobulin antibodies. I. Mode of action. Eur. J. Immunol. 4, 715 (1974).

(10) Melchers, F., von Boehmer, H. & Phillips, R.A. B-lymphocytes subpopulations in the mouse. Organ distribution and ontogeny of immunoglobulin-synthesizing and of mitogen-sensitive cells. Transplant. Rev. 25, 26 (1975).

(11) Melchers, F., Andersson, J., & Phillips, R.A. Ontogeny of Murine B-Lymphocytes: Development of Ig synthesis and of reactivities to mitogens and to anti Ig antibodies. Cold Spring Harbor Symp. Quant. Biol. 41, 147 (1977).

(12) Pernis, B., Forni, L. & Luzzati, A.L. Synthesis of multiple immunoglobulin classes by single lymphocytes. Cold Spring Harbor Symp. Quant. Biol. 41, 175 (1977).

(13) Vitetta, E.S., Cambier, J., Forman, J., Kettman, J.R., Yuan, D. & Uhr, J.W. Immunoglobulin receptors on murine B-lymphocytes. Cold Spring Harbor Symp. Quant. Biol. 41, 185 (1977).

(14) Parkhouse, R.M.E., Abney, E.R., Bourgois, A. & Willcox, H.N.A. Functional and structural characterization of immunoglobulin on murine B-lymphocytes. Cold Spring Harbor Symp. Quant. Biol. 41, 93 (1977).

(15) Andersson, J., Coutinho, A. & Melchers, F. Frequencies of mitogen-reactive B-cells in the Mouse. I. Distribution in different lymphoid organs from different inbred strains of mice at different ages. J. exp. Med. 145, 1511 (1977).

(16) Andersson, J., Coutinho, A. & Melchers, F. The switch from IgM- to IgG-Secretion in single clones of mitogen-activated B-cells. J. exp. Med. (in press).

(17) Andersson, J., Coutinho, A., Lernhardt, W. & Melchers, F. Clonal growth and maturation to immunoglobulin secretion in vitro of every growth-inducible B-lymphocyte. Cell 10, 27, (1977).

(18) Potter, M. The plasma cell tumors and myeloma proteins of mice. In: Methods in Cancer Research, ed. Busch, H., Vol. 2, p. 105, Academic Press, New York.

(19) Andersson, J., Buxbaum, J., Citronbaum, R., Douglas, S., Forni, L., Melchers, F., Pernis, B. & Stott, D. IgM producing tumors in the Balb/C mouse. A model for B-cell maturation. J. exp. Med. 140, 742 (1974c).

(20) Milstein, C., Brownlee, G., Harrison, T.M. & Mathews, M.B. A possible precursor of immunoglobulin light chains. Nature New Biol. 239, 117 (1972).

(21) Swan, D., Aviv, H. & Leder, P. Purification and properties of biologically active messenger RNA for a myeloma light chain. Proc. Nat. Acad. Sci. USA 69, 1967 (1972).

(22) Burstein, Y. & Schechter, I. Amino acid variability at the N-terminal extra piece of mouse immunoglobulin light chain-precursors of the same and different subgroups. Biochem. J. 157, 145 (1977).

(23) Burstein, Y. & Schechter, I. Amino Acid sequence of the $NH_2$-terminal extra piece segments of the precursors of mouse immunoglobulin λI-type and κ-type light chains. Proc. Nat. Acad. Sci. USA 74, 716 (1977).

(24) Blobel, G. & Dobberstein, B. Transfer of Proteins Across Membranes. I. Presence of proteolytically processed and unprocessed nascent immunoglobulin light chains on membrane-bound ribosomes of Murine Myeloma. J. Cell. Biol. 67, 835 (1975).

(25) Blobel, G. & Dobberstein, B. Transfer of proteins across membranes. II. Reconstitution of functional rough microsomes from heterologous conponents. J. Cell. Biol. 67, 852 (1975).
(26) Melchers, F. Biosynthesis of carbohydrate portion of immunoglobulins. Kinetics of synthesis and secretion of [$^{3}$H]-mannose-labeled myeloma protein by two plasma cell tumors. Biochem. J. 119, 765 (1970).
(27) Melchers, F. Biosynthesis of the carbohydrate portion of immunoglobulin. Radiochemical and chemical analysis of the carbohydrate moieties of two myeloma proteins purified from different subcellular fractions of plasma cells. Biochemistry 10, 653 (1971).
(28) Melchers, F. Difference in carbohydrate composition and a possible conformational difference between intracellular and extracellular immunoglobulin M. Biochemistry 11, 2204 (1972).
(29) Melchers, F. Biosynthesis, intracellular transport and secretion of immunoglobulins. Effect of 2-deoxy-D-glucose in tumor plasma cells producing and secreting immunoglobulin Gl. Biochemistry 12, 1471 (1973).
(30) Melchers, F. & Andersson, J. Synthesis, surface deposition and secretion of immunoglobulin M in bone-marrow-derived lymphocytes before and after mitogenic stimulation. Transplant. Rev. 14, 76 (1973).
(31) Scharff, M.D. & Laskov, R. Synthesis and assembly of immunoglobulins. Progr. Allergy 14, 37 (1970).
(32) Parkhouse, R.M.E. & Askonas, B.A. Immunoglobulin M biosynthesis. Intracellular accumulation of 7 S subunits. Biochem. J. 115, 163 (1969).
(33) Baenziger, J. & Kornfeld, S. Structure of the carbohydrate units of $IgA_1$ immunoglobulin. J. Biol. Chem. 249, 7270 (1974).
(34) Parkhouse, R.M.E. & Melchers, F. Biosynthesis of the carbohydrate portions of immunoglobulin M. Biochem. J. 125, 235 (1971).
(35) Hsu, A.-F., Baynes, J.W. & Heath, E.C. The role of a dolichol oligosaccharide as an intermediate in glycoprotein biosynthesis. Proc. Natl. Acad. Sci. USA 71, 2391 (1974).
(36) Eylar, E.H. On the biological role of glycoproteins. J. theor. Biol. 10, 89 (1965).
(37) Melchers, F. & Knopf, P.M. Biosynthesis of the carbohydrate portion of immunoglobulin chains: possible relation to secretion. Cold Spring Harbor Symp. Quant. Biol. 32, 255 (1967).
(38) Farkas, V., Svoboda, A. & Bauer, S. Secretion of cell-wall glycoproteins by yeast protoplasts. Effect of 2-deoxy-D-glucose and cycloheximide. Biochem. J. 118, 755.
(39) Liras, P. & Gascon, S. Biosynthesis and secretion of yeast invertase: Effect of cycloheximide and 2-deoxy-D-glucose. Europ. J. Biochem. 23, 160 (1971).
(40) Gandhi, S.S., Stanley, P., Taylor, J.M. & White, D.O. Inhibition of influenza viral glycoprotein synthesis by sugars. Microbios.

5, 41 (1972).

(41) Tkacz, J.S. & Lampen, J.O. Tunicamycin inhibition of polyisoprenyl-N-acetylglucosaminyl pyrophosphate formation in calf-liver microsomes. Biochem. Biophys. Res. Commun. 65, 248 (1975).

(42) Takatsuki, A., Kohno, K. & Tamura, G. Inhibition of biosynthesis of polyisoprenol sugars in chick embryo microsomes by Tunicamycin. Agri. Biol. Chem. 39, 2089 (1976).

(43) Lehle, L. & Tanner, W. The specific site of Tunicamycin inhibition in the formation of dolichol-bound N-acetylglucosamine derivatives. FEBS Lett. 71, 167 (1976).

(44) Hickman, S., Kulczycki, A., Lynch, R.G. & Kornfeld, S. Studies of the mechanism of tunicamycin inhibition of IgA and IgE secretion by plasma cells. J. Biol. Chem. 252, 4402 (1977).

(45) Hickman, S. & Kornfeld, S. Effect of Tunicamycin on IgM, IgA and IgG Secretion by mouse plasmacytoma cells. J. Immunol. in press, (1978).

(46) Melchers, F. The secretion of a Bence-Jones type light chain from a mouse plasmacytoma. Europ. J. Immunol. 1,330 (1971).

(47) Weitzman, S. & Scharff, M.D. Mouse myeloma mutants blocked in the assembly, glycosylation and secretion of immunoglobulin. J. Mol. Biol. 102, 237 (1976).

(48) Andersson, J., Lafleur, L. & Melchers, F. Immunoglobulin M in bone marrow-derived lymphocytes. Synthesis surface deposition turnover and carbohydrate composition in unstimulated mouse B-cells. Eur. J. Immunol. 4, 170 (1974).

(49) Melchers, F. & Andersson, J. Immunoglobulin M in bone marrow-derived lymphocytes. Changes in synthesis, turnover and secretion, and in number of molecules on the surface of B-cells after mitogenic stimulation. Eur. J. Immunol. 4, 181 (1974).

(50) Melchers, F. & Andersson, J. Early changes in immunoglobulin M synthesis after mitogenic stimulation of bone marrow-derived lymphocytes. Biochemistry 13, 4645 (1974).

# IRON SULFUR CENTERS RECONSTITUTIVE CAPACITY: DECAY AND RESTORING IN SUCCINATE DEHYDROGENASE

F.Bonomi, S.Pagani and P.Cerletti, Dept. of General Biochemistry, University of Milan, I 20133, Milano, Italy.

In recent years the EPR studies on succinate dehydrogenase (succinate: acceptor oxidoreductase E.C.1.3.99.1) made clear that presence of the HiPIP type signal in soluble preparations is closely related to the reconstitutive capacity of the preparation (1, 2): it was also established that the latter property decays in parallel with the catalytic activity at low ferricyanide concentration (3). It has therefore been questioned whether all three parameters depend on the state of a same structure, namely the 4Fe-4S iron sulfur center of the flavoprotein, or whether reconstitutive capacity and catalytic activity at low ferricyanide reflect the integrity of a particular region of the enzyme which includes also the 4Fe-4S center.

Methods that allow to remove and insert selectively sulfide and iron in the flavoprotein are an approach to investigate the state of iron sulfur centers of the dehydrogenase and its implications on other properties of the enzyme. It is at present clear that losses in protein bound labile sulfide are not necessary to destroy the reconstitutive capacity since flavoprotein preparations have been described with full complement in labile sulfide which are reconstitutively inert (4). Iron and sulfide can be chemically introduced into the flavoprotein restoring its reconstitutive capacity (4) and we showed that the sulfurtransferase rhodanese (E.C.2.8.1.1) inserts sulfide sulfur into the flavoprotein thereby protecting and implementing the reconstitutive capacity of the latter. Optical and EPR spectra indicate that the iron sulfur structures are involved in the process (5, 6).

The effect of $^{35}S$ incorporation on labile sulfide content parallels but is less than that on reconstitutive capacity (fig. 1). This latter property therefore appears related to labile sulfide but at least in part it does not depend on a net increase of it. The nega-

---

Abbreviations used are: BSA: bovine serum albumin; PCMS: p-chloromercury-sulfonate; PMS: phenazine methosulfate.

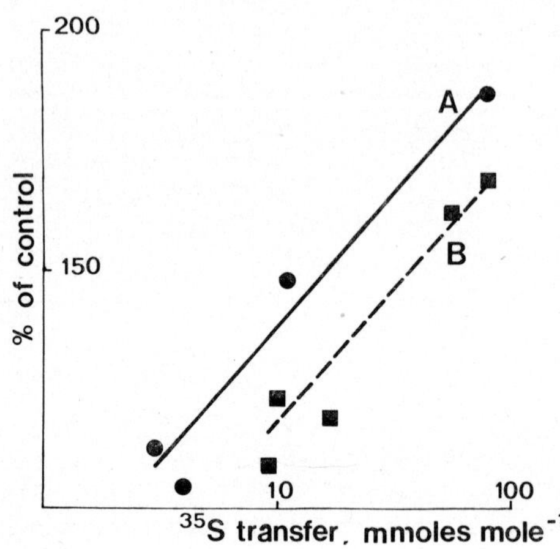


Fig. 1. Rhodanese mediated sulfur incorporation in succinate dehydrogenase and protection of reconstitutive activity and of labile sulfide content. Succinate dehydrogenase was incubated 60 min at 0 C without and with rhodanese labelled with $^{35}S$ in its persulfide sulfur, added in a 1:3 molar ratio to FAD. For each experiment the reconstitutive capacity (A) the labile sulfide content (B) were measured at the end of incubation and the amount of $^{35}S$ sulfur incorporated was determined after isolating the flavoprotein. Abscissae give this latter quantity: ordinates give A and B in the sample with rhodanese as percent of the sample without rhodanese. Lines are least square fittings.

tive intercepts on the ordinates indicate that the interaction with rhodanese implies reshuffling the iron sulfur structure (more, and the molecular region responsible for reconstitution (less), with some inherent damage. On the other hand as shown in table 1, rhodanese does not affect the catalytic activity with ordinary acceptors (phenazine methosulfate), but it does with ferricyanide at low concentrations, which confirms the relations of this assay to reconstitutive capacity.

In the interaction with rhodanese, sulfur originates from a sulfane donor, e.g. thiosulfate, and is transferred _via_ the active persulfide group of the transferase; it is unsettled whether its reduction to sulfide occurs on the transferase or on the flavoprotein. Both the oxidized dehydrogenase and the enzyme reduced by succinate act as acceptors, but sulfide incorporated in oxidizing conditions is more labile and decays to the medium (5, 6).

TABLE 1

EFFECT OF RHODANESE ON THE CATALYTIC ASSAYS AND RECONSTITUTIVE CAPACITY OF SUCCINATE DEHYDROGENASE

Succinate dehydrogenase (8.2 nmoles histidyl FAD (mg protein)$^{-1}$) was incubated 60 min at 0 C in a nitrogen atmosphere without or with rhodanese added in a 1:3 molar ratio to FAD. Activities with phenazine methosulfate and with low concentration of ferricyanide and the reconstitutive capacity were assayed before and after incubation. Preparations and assays here and in following tables and figures were as described in ref. 5. The activity at low ferricyanide concentration was determined according to ref. 3.

| | Catalytic activity µmoles succinate min$^{-1}$(mg protein)$^{-1}$ | | | | Reconstitutive capacity, % | |
|---|---|---|---|---|---|---|
| | PMS | | (150 µM) $Fe(CN)_6^{3-}$ | | | |
| additions: min at 0 C | none | rhodanese | none | rhodanese | none | rhodanese |
| 0 | 13.00 | - | 1.30 | - | 20.5 | 22.5 |
| 60 | 12.30 | 12.20 | 0.86 | 0.98 | 18.8 | 40.6 |

Similar events have been observed with other iron sulfur proteins, namely spinach ferredoxin which has one 2Fe-2S center and the clostridial ferredoxin, having two 4Fe-4S clusters. The amount transferred waries with the acceptor and probably sulfur is reduced by different mechanisms (7).

## Iron sulfur centers and reconstitution

Agents which supposedly may affect the iron sulfur structure of succinate dehydrogenase are shown for their effect on the reconstitutive capacity in table 2.

The reactants for chemically restoring the reconstitutive capacity had little effect as compared e.g. to the known protection by rhodanese: this does not surprise since the original procedure uses a higher temperature (4). The situation changed completely when iron and sulfide were carried into the reaction bound to a protein. The artificial iron sulfur protein synthesyzed from bovine serum albumin contained per mole 7.12 moles sulfide bound in a labile form; its

TABLE 2

EFFECT OF SULFUR DONORS ON THE RECONSTITUTIVE CAPACITY OF SUCCINATE DEHYDROGENASE

Succinate dehydrogenase (7.3 nmoles histidyl FAD $mg^{-1}$) was incubated 60 min at 0°C under nitrogen with the compounds indicated at the given concentrations or molar ratios (x:FAD). The reconstitutive capacity of the flavoprotein was then assayed.

| Added | Reconst. capacity % | Added | Reconst. capacity % |
|---|---|---|---|
| 0 | 20.6 | BSA 1:3 | 23.3 |
| rhodanese 1:3 | 32.1 | iron-sulfur BSA 1:3 | 32.1 |
| 4mM $Na_2S$ + 4mM $Fe(NH_3)_2(SO_4)_2$ + 5mM dithiothreitol | 23.5 | same 3:3 | 35.5 |
| same + rhodanese 1:3 | 31.2 | clostridial ferredoxin 1:3 | 27.5 |
| spinach ferredoxin 1:3 | 19.5 | same 3:3 | 31.6 |
| same 3:3 | 24.4 | same + rhodanese 1:3:1 | 29.8 |

optical spectra matched those of spinach ferredoxin. As shown in table 2, it protected succinate dehydrogenase even better than rhodanese and the effect depended on the ratio between the two interacting proteins: since BSA as such did not produce it, it should be attibuted to the iron sulfur component associated to the albumin.

Similar results were obtained with clostridial ferredoxin. The quantitatively smaller effect may be due to the decayed state of the preparation used which contained 2.61 moles sulfide per mole instead of eight, such that the molar ratio of exogenous sulfide to flavin was less than in the experiments with iron-sulfur BSA. Rhodanese did not interfere with the interaction.

Spinach ferredoxin was much less effective; the preparation contained 1.92 moles sulfide per mole: the ratio of sulfide to FAD being comparable with that in the experiments with the clostridial protein, the small effect measured is intrinsic to this type of ferredoxin. It is interesting that clostridial ferredoxin which contains 4Fe-4S centers is more effective than spinach ferredoxin in protecting the reconstitutive capacity of the flavoprotein since it is known that this property decays in parallel with the HiPIP type signal from the 4Fe-4S center of the flavoprotein.

In summary protein bound iron and sulfide were more effective than the inorganic counterparts, with a probable preference for one type of cluster. Presence of iron and sulfide in an organized structure may favour incorporation: indeed with synthetic analogues and also with plant and bacterial ferredoxin extrusion and replacement of the iron sulfide core of 2Fe-2S and 4Fe-4S clusters has been produced, and interconversion of dimeric and tetrameric structures has been observed in synthetic analogues (8).

All donors considered, except rhodanese, contained sulfur at the oxidation level of sulfide, and protected sametimes better than rhodanese itself. This indicates that succinate dehydrogenase accepts easily sulfide sulfur and confirms our belief that with this protein sulfur from rhodanese is transferred as sulfide (6). The reducing equivalents necessary to bring to this state the persulfide sulfur at the active site of the transferase come from its cysteines, as shows the measured decrease of thiols in rhodanese reacted with succinate dehydrogenase (6). If the disulfide formed involves the SH group at the active site of the enzyme, the transferase is inactivated; when thiosulfate is present and turns over with rhodanese it produces sulfi e which may regenerate by sulfitolysis of the disulfide one free SH group: if this is the catalytic thiol, the transferase is reactivated since presence of one sulfite residue on rhodanese does not inactivate it (Dr. Cannella, personal communication). Thiosulfate is much less effective if it is added to the inactivated transferase after incubation with the flavoprotein: likely, if it is present during the reaction with the dehydrogenase it binds to the transferase molecule and prevents irreversible conformational changes connected with disulfide formation.

In the hypothesis outlined sulfide formation occurs on the rhodanese molecule as an independent process from sulfane transfer to the acceptor protein. The alternative hypotesis that sulfide is generated by an intramolecular event in succinate dehydrogenase is less likely since cysteines necessary for iron sulfur clusters may be involved, and, if so, full restoral of the iron sulfur structure becomes impossible.

Membrane bound succinate dehydrogenase being much more stable than the soluble counterpart, effects on spontaneous decay cannot be evi-

TABLE 3

EFFECT OF RHODANESE ON PARTICULATE SUCCINATE DEHYDROGENASE

Keilin Hartree heart muscle preparation (initial activity 0.209 µmoles $O_2$ oxidized min (mg protein)$^{-1}$) was incubated 60 min at 37°C and pH 9.5 without and with rhodanese (0.263 mole per mole FAD) and 1 mM thiosulfate added. The residual succinooxidase activity was then measured. An aliquot of sample incubated without rhodanese (ATKH) was incubated further 60 min with rhodanese at 0°C and the activity assayed.

| Additions | Residual activity % |
|---|---|
| none | 0 |
| rhodanese | 13.4 |
| rhodanese + 1mM $Na_2S_2O_3$ | 9.6 |
| rhodanese added to ATKH | 0 |

denced. We therefore incubated it in alkaline medium; it is known that this treatment modifies the flavoprotein but does not detach it from the membrane (9); the inactivated enzyme does not further interact with the terminal oxidase system, though this remains effective as shown by restored activity after adding reconstitutively active soluble succinate dehydrogenase. The succinooxidase activity of the particulate enzyme measures its fitness to interact with the subsequent carriers in the respiratory system, that means the same molecular condition evidenced by reconstitutive capacity in the soluble flavoprotein. We observed that rhodanese, with or without thiosulfate present, decreased the inactivation of the particulate dehydrogenase but did not reactivate the membrane bound enzyme after it was inactivated (table 3); it is possible that binding of the flavoprotein to the membrane may hinder molecular modifications connected with repair of iron-sulfur structures.

## The action of mercurials

The effects of mercurials on EPR signal intensity, on the reconstitutive capacity (measured by activity at low ferricyanide concentration) and on catalytic activity of succinate dehydrogenase towards phenazine methosulfate are given in table 4. The shape of signals is shown in figure 2.

In anaerobiotic aging the signal of center S-1 decayed more rapidily than the one elicited by dithionite (centers S-1 + S-2), S-3 having an intermediate behaviour. The decreased signal from center S-1 with aging may in part be due to decreased efficiency of reduction with succinate since the peak height of the signal with dithionite was affected only slightly.

The catalytic activity with phenazine methosulfate was more stable to aging than the one measured at low concentration of ferricyanide, which decays in parallel with the reconstitutive capacity (3). The reactivity at the catalytic site for succinate was the same with either dye, as indicated by equal percent inhibition by PCMS: it is known that mercurials have high affinity for the thiol group at the active site of the dehydrogenase (10); therefore the larger decay in the activity measured with ferricyanide was representative of changes in other molecular parameters.

In presence of mercurials the reconstitutive capacity, indicated by the activity at low ferricyanide concentration, decayed with different kinetics than iron sulfur structures, represented by EPR signals: this indicates that the former involves other molecular parameters beside the state of these clusters.

Due to their high affinity for the catalytic SH of succinate dehydrogenase mercurials at stoichiometric concentration with free SH groups in the flavoprotein (9 per mole (11)) completely blocked, and at one tenth this concentration significantly diminished the reduction by succinate of dye and of center S-1 whereas reduction by dithionite was practivally unaffected; a 9:1 ratio of PCMS to FAD modified only slightly the signals elicited by dithionite from centers S-1 and S-2; these structures became totally inert only with excess mercurial. Center S-3 does not participate of the effects at the catalytic SH and appears less affected by mercurials than the two other clusters. However in soluble succinate dehydrogenase the HiPIP type signal accounts for only part of the flavin in the preparation (1, 2) and decays more in aging: a quantitative appraisal is not possible since some structures may be damaged by mercurials but some other may become responsive.

The EPR spectra of succinate added mercurial-treated dehydrogenase in which the signal at g = 1.94 is heavily decreased or is absent, show practically unmodified a signal at g = 2.03: it combines the $g_z$ component of the signal of center S-1, disappearing because of the action of mercurials, and the HiPIP type signal of that part of enzyme which is not reduced due to PCMS blocking the catalytic SH. The signal of the flavin semiquinone at g = 2.00 disappears in parallel with one at g = 1.94.

TABLE 4

EFFECT OF MERCURIALS ON CATALYTIC ACTIVITY AND EPR SIGNALS OF SDH

Succinate dehydrogenase was aged anaerobically at 0°C. At the time indicated p-chloromercury--sulfonate was added to part of the sample. Conditions for incubation and EPR assay are those described in ref. 3 except for omission of succinate during aging. Enzymic activities and peak height of EPR signals of samples which did not receive PCMS are given as percent of the value before incubation. Values in the PCMS treated sample are given as the percent of the untreated sample.

| time min | PCMS added mol/mol FAD | SDH activity | | EPR signals: peak height | | |
|---|---|---|---|---|---|---|
| | | | | $g = 1.94$ | | $g = 2.03$ |
| | | 150 µM $Fe(CN)_6^{3-}$ | PMS | succinate reduced | dithionite reduced | 150 µM $Fe(CN)_6^{3-}$ |
| 0 | 0 | 100 | 100 | 100 | 100 | 100 |
| 30 | 0 | 57.0 | 62.2 | 69.3 | 96.5 | 88.6 |
| | 180 | (1.9) | (0.01) | 0 | 0 | 0 |
| 50 | 0 | 34.0 | 54.9 | 54.3 | 94.2 | 81.8 |
| | 9 | (2.8) | (4.2) | 0 | (82.3) | (94.5) |
| 80 | 0 | 12.0 | 22.1 | 37.7 | 91.0 | 72.5 |
| | 0.9 | (35.1) | (39.6) | (82.8) | (96.5) | (115.1) |

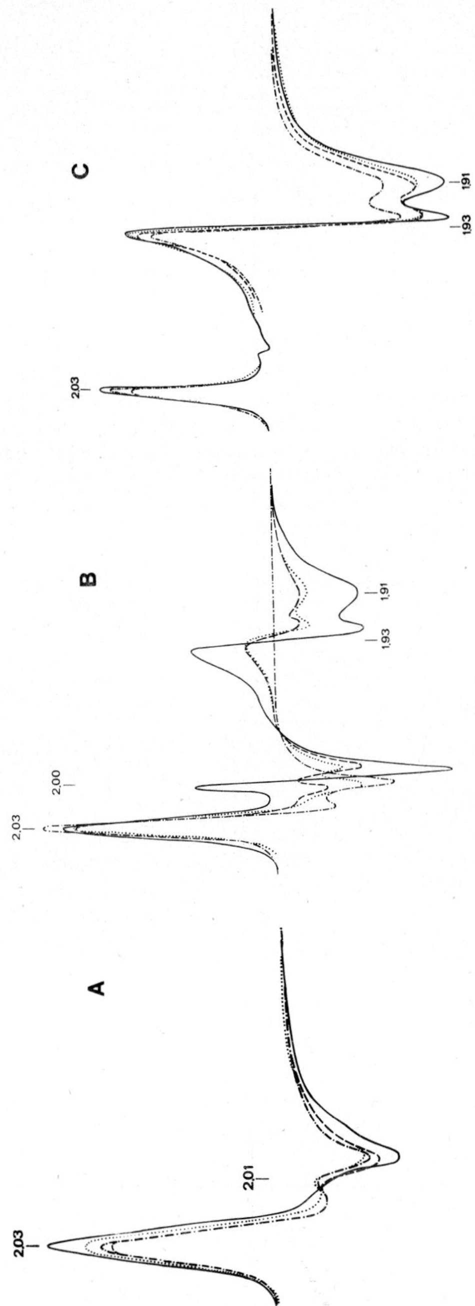


Fig. 2. EPR spectra of succinate dehydrogenase treated or not with mercurials. Measured samples are described in table 4. Full line: before aging; dots: aged 80 min at 0 C. Dashes: aged 80 min than PCMS added 0.9 molar to histidyl FAD; dashes and dots: aged 50 min than PCMS added 9 molar to histidyl FAD. A: sample oxidized with 140 µM ferricyanide and 10 µM phenazine methosulphate. B: sample reduced with 20 mM succinate. C: sample reduced with excess dithionite.

## Concluding remarks

The type of action of mercurials and the behaviour of reconstitutive capacity and of labile sulfide content as a function of sulfur incorporation into the flavoprotein indicate that the reconstitutive capacity is certainly influenced by the state of iron sulfur centers but depends also on other molecular parameters.

Rhodanese acts primarily through insertion of sulfide but the net effect appears more pronounced for the reconstitutive capacity: this suggests that incorporation of exogenous sulfur is a dynamic process which involves degrading and restoring the existing iron sulfur structures concomitant to other molecular modifications. Reduction of sulfane sulfur to sulfide on the rhodanese molecule, which may inactivate the transferase, is considered as occurring only in the *in vitro* system: in the cell other redox systems are available which may easily couple with the reduction of the persulfide sulfur of the enzyme.

Insertion of sulfur by rhodanese is a probable way for cellular synthesis of iron sulfur structures. Nevertheless the remarkable effect of protein bound iron and sulfide on the reconstitutive capacity of succinate dehydrogenase suggests that iron sulfur proteins in a same cell compartment may interact to build anew or to restore their iron sulfur centers.

## SUMMARY

Agents such as aging, rhodanese, mercurials, modified both the reconstitutive capacity and the state of iron sulfur centers in succinate dehydrogenase. These two parameters though they varied in parallel, they do not appear to depend on a same molecular component. The effect on the dehydrogenase of inorganic and of protein bound iron and sulfide has also been studied.

## ACKNOWLEDGMENT

This investigation was supported in part by a grant of the Italian National Research Council (C.N.R.).

EPR measurements were performed at the Fachbereich Biologie of the University of Konstanz by Dr. P.Kroneck to whom grateful acknowledgement is made.

## REFERENCES


1) Ohnishi, T., Lim, J., Winter, D.B., and King, T.E. (1976) J.Biol. Chem. 251, 2105-2109.

2) Beinert, H., Ackrell, B.A.C., Vinogradov, A.D., Kearney, E.B., and Singer, T.P. (1977) Arch.Biochem.Biophys. 182, 95-106.

3) Vinogradov, A.D., Ackrell, D.A.C., and Singer, T.P. (1975) Biochem.Biophys.Res.Commun. 67, 803-809.

4) Baginsky, M.L., and Hatefi, Y. (1969) J.Biol.Chem. 244, 5313-5319.

5) Bonomi, F., Pagani, S., Cerletti, P., and Cannella, C. (1977) Eur. J.Biochem. 72, 17-24.

6) Bonomi, F., Pagani, S., and Cerletti, P. (in press) Flavins and Flavoproteins (Yagi, K., and Yamano, T., eds), Scientific Society Press, Tokyo.

7) Bonomi, F., Pagani, S., and Cerletti, P. (1977) FEBS Letters 84, 149-152.

8) Holm, R.M., and Ibers, J.A. (1977) in Iron-sulfur Proteins, vol. III (Lovenberg, A., ed), Academic Press, Inc. 205-281.

9) Hanstein, W.G., Davis, K.A., Ghalambor, M.A., and Hatefi, Y.(1971) Biochemistry 10, 2517-2524.

10) Vinogradov, A.D., and Zuevsky, V.V. (1973) FEBS Letters 36, 99-101.

11) Pagani, S., Bonomi, F., Cerletti, P. (1974) FEBS Letters 39, 139-143.

SESSION II

# Selective Degradation of Proteins

# METHODS FOR DETERMINING PROTEIN TURNOVER

P.J. Garlick, E.B. Fern and M.A. McNurlan
Department of Human Nutrition,
London School of Hygiene and Tropical Medicine,
Keppel Street, London WC1E 7HT.

The conventional method for measuring the rate of protein breakdown *in vivo* involves injection of a labelled protein precursor and measurement of the rate of decay of labelled protein. The major difficulty, that of recycling of label for synthesis of new protein, can generally be overcome by a suitable choice of non-reutilizable label, e.g. $NaH^{14}CO_3$ (1,2), $[6^{14}C]$arginine (3), $[^3H]$5 amino-levulinic acid (4). The method relies on the principle that, if label is lost randomly from the protein, the decay will be exponential. By plotting the decay curve on semi-log axes, a straight line is obtained whose gradient is the fractional rate of protein breakdown. With single, purified proteins this simple technique has been used with good success (e.g. 3). The method is not, however, useful for measuring the mean degradation rate of protein mixtures during studies of the growth and regulation of tissue protein mass, since decay is no longer exponential. Therefore we shall describe, firstly a different method of analyzing the decay curve and secondly an alternative method for estimating breakdown by measuring the rate of synthesis and the rate of change of protein mass.

## ESTIMATES OF BREAKDOWN OF WHOLE TISSUE PROTEINS FROM DECAY CURVES

A typical decay curve for liver protein after labelling with $NaH^{14}CO_3$ is shown in Fig. 1 (2,5). It is apparent that, even when plotted on semi-log axes, it is not a straight line. The frequently used method of making measurements at only two points in time, and drawing a straight line between them, cannot therefore give an unambiguous value for the rate of breakdown. For example, the

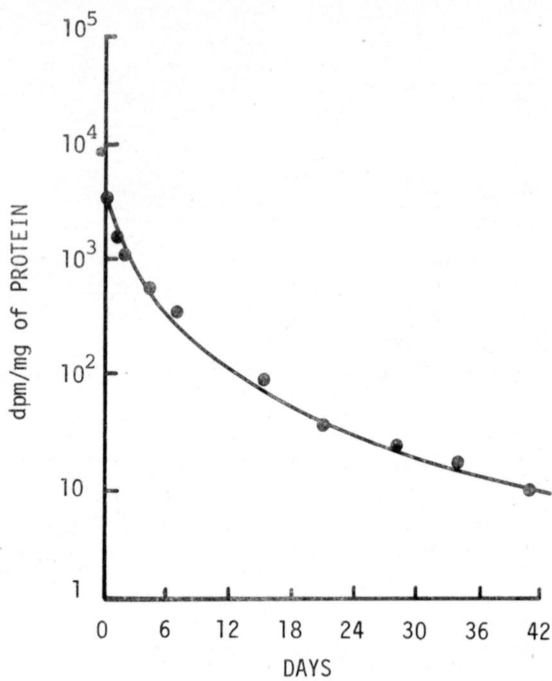


Fig. 1. The specific activity of liver protein in rats at different times after injection of $NaH^{14}CO_3$. Female rats (240 g) were given intraperitoneal injections of 1 mCi $Na_2{}^{14}CO_3$ and groups of 4 killed at intervals of 1 hour to 41 days thereafter (2,5).

line drawn between peak labelling (1 h) and 1 day indicates that 77% of proteins are broken down each day ($T_{\frac{1}{2}}$ = 0.9 days), whereas that between 1 day and 30 days gives a rate of 15%/day ($T_{\frac{1}{2}}$ = 4.7 days). The reason for this is that a mixture of proteins does not behave as if it were a single protein with a breakdown rate equal to the mean rate for the mixture. An analysis which takes account of the heterogeneity of turnover rates in the mixture must be used.

## Multi-exponential analysis

The curve shown in Fig. 1 has been separated into 3 exponential components (5). This implies that the mixture of proteins in liver can be regarded as 3 groups with fast, medium and slow turnover rates.

TABLE 1 Turnover Rates of 3 Components of Liver Protein Obtained by 3 Exponential Analysis of the Decay Curve

| Component | Mass (%) | Turnover rate (%/d) |
|---|---|---|
| 1 | 9.5 | 228 |
| 2 | 66.4 | 27 |
| 3 | 24.1 | 6.2 |

The components were obtained by separating the decay curve of liver protein (Fig. 1) into 3 exponentials, i.e.

$$\text{radioactivity} = X_1 e^{-\lambda_1 t} + X_2 e^{-\lambda_2 t} + X_3 e^{-\lambda_3 t}.$$

The fractional breakdown rates of the 3 components were taken to be equal to the 3 values of $\lambda$ and their relative masses were given by 3 values of $X_i/\lambda_i$ (5).

The masses and turnover rates of these 3 components have been calculated from the rate constants and coefficients of the 3 exponentials and are shown in Table 1. It is interesting that in spite of the large range of turnover rates for liver enzymes which have been studied individually (6,7), the majority of protein seems to fall into component 2, which has a turnover rate close to the mean for the total mixture.

From the masses and turnover rates of the 3 components shown in Table 1 it is possible to calculate the mean rate of turnover of total liver protein (41%/day, $T_{\frac{1}{2}}$ = 1.7 days). The mean rate can also be calculated from the area under the specific activity-time curve as described by Garlick et al (5). These analyses, however, require a decay curve to be measured over at least 20 days if reasonably accurate estimates of turnover rates are to be obtained. There are very few studies in which this would be practical since the rate of degradation is usually of interest during changes in protein mass which take only a few days to complete.

## Approximate method for mean turnover rate

Garlick et al (5) showed that a straight line drawn between the point of maximal labelling and the point when the radioactivity has fallen to 10% of this maximum has a gradient approximately equal to the mean fractional rate of degradation. The reason for this relationship is unknown, but it gives the correct result both with theoretical model mixtures of protein and with the decay curve shown in Fig. 1. The time period necessary for decay to 10% of peak labelling is between 3 to 4 times the mean half-life for the mixture; that is about 6 days for liver. This time period is much shorter than is necessary for the multi-exponential analysis, but is still unacceptably long. For tissues whose rates of turnover are slower than liver, for example muscle, even this method might take more than 20 days to perform.

We conclude this discussion of the decay curve method by pointing out that, even when non-reutilizable precursors are used to label protein, the method is not very suitable for measurements of the breakdown rate of whole tissue proteins. The decay curve must be analysed in a way which takes into account the heterogeneity of turnover rates in the mixture. This can give useful information on the distribution of turnover rates in the tissue and has been used by Scornik and Botbol (8) to show that the reduced rate of protein breakdown after partial hepatectomy affects fast, medium and slowly turning over components equally. However, even when the shorter, approximate method is used to obtain the mean rate of turnover, the method requires that the decay of label be followed for a prohibitively long time. In most types of experiment, therefore, other methods of measuring breakdown have to be found.

## ESTIMATES OF BREAKDOWN FROM MEASUREMENTS OF SYNTHESIS AND GROWTH

Measurement of the rate of synthesis from the incorporation of labelled amino acids into protein is not affected by the heterogeneity of turnover rates in whole tissues so long as incorporation is measured over a relatively short period lasting not more than a few hours (5,9). It seems logical, therefore, that breakdown should be estimated indirectly as the difference between the measured rate of protein synthesis and the net rate of growth of protein mass. Hence the fractional rate of degradation ($k_d$) is given by the expression, $k_d = k_s - k_g$ . $k_s$ is the fractional rate of synthesis (fraction or percent renewed per day) and $k_g$ is the fractional rate of growth (rate of accumulation of protein divided by mass of protein already present). We have found this indirect method of assessing breakdown to be particularly valuable for muscle because of the difficulty of obtaining valid measurements by the decay curve method. Examples of the use of this method for determining muscle protein breakdown are given below. Similar methods have been used for liver proteins by Scornik and coworkers (8,10).

### Measurement of synthesis rate

We have generally estimated the rate of protein synthesis in muscle from the incorporation of label into protein during constant intravenous infusion of [$^{14}$C]tyrosine (9,11,12,13). The label is given as an infusion, rather than as a single injection, because the free amino acid specific activity in plasma and tissues rapidly becomes constant (plateau). Allowance is made for the short time before plateau specific activity is reached by assuming that the specific activity in the tissue rises to plateau by a single exponential pathway. The rate constant of this exponential in muscle is related to the rate of protein turnover and the ratio of amounts of protein bound to free amino acid in that tissue. There is then no need to determine the time course of free amino acid specific activity, as is necessary with single injection methods by making measurements at several time points (e.g. 14,15). The fractional rate of protein synthesis ($k_s$) can be calculated from the specific activity of the free and protein bound amino acid at the end of infusion.

## The Response to a Protein-free Diet

When a protein-free diet was given to young (100 g) rats there was a loss of body weight (16) and a very slow decline in the protein mass of gastrocnemius muscle (Table 2). On the first day there was a drop in protein synthesis to half the rate in normally growing control animals. Thereafter, for the following 20 days, there was a slow decline in synthesis to about 25% of its initial rate. These changes were accompanied by a 60% fall in the RNA content of the muscle (16). Low rates of protein synthesis in muscle of rats given a protein-free diet have also been demonstrated by Waterlow and Stephen (11), who estimated synthesis by constant infusion of [$^{14}$C] lysine. The calculated rate of protein breakdown also fell on the first day of the diet. For the remaining 20 days the rates of syn-

TABLE 2 The Effect of a Protein-Free Diet on Muscle Protein Turnover Measured by Constant Infusion of [$^{14}$C]tyrosine

| Days on diet | Protein mass (% of control ± SEM) | $k_s$ (%/d ± SEM) | $k_d$ (%/d) |
|---|---|---|---|
| 0 | 100 ± 2 | 18.7 ± 0.8 | 13.2 |
| 1 | 97 ± 2 | 9.7 ± 0.4 | 9.7 |
| 2 | 103 ± 3 | 8.7 ± 0.7 | 8.7 |
| 3 | 103 ± 3 | 8.2 ± 0.7 | 8.2 |
| 9 | 95 ± 4 | 5.1 ± 0.6 | 6.5 |
| 21 | 87 ± 3 | 4.3 + 0.6 | 5.7 |

All rats were given a diet containing 12% casein before transfering to the protein free diet at a body weight of about 110 g (16).

TABLE 3 The Effect of Starvation on Protein Turnover in Muscle: Measurement of Synthesis by Constant Infusion of $[^{14}C]$Tyrosine

| Days of Starvation | Protein mass (% of control $\pm$ SEM) | $k_s$ (%/d $\pm$ SEM) | $k_d$ (%/d) |
|---|---|---|---|
| 0 | 100 $\pm$ 3 | 16.5 $\pm$ 1.7 | 8.9 |
| 1 | 101 $\pm$ 4 | 8.5 $\pm$ 1.4 | 8.5 |
| 2 | 100 $\pm$ 3 | 6.9 $\pm$ 0.9 | 6.9 |
| 3 | 88 $\pm$ 3 | 4.6 $\pm$ 1.6 | 14.6 |
| 4 | 76 $\pm$ 8 | 4.7 $\pm$ 1.6 | 18.7 |

Rats weighing about 120 g were fed a commercial cube diet before commencing starvation. Data of Millward (19, 20).

thesis and breakdown decreased in an almost parallel manner, as indicated by no measurable loss of protein until after 3 days, and the very slow loss subsequently. Independent evidence for a fall in muscle protein degradation comes from measurements of the urinary excretion of 3 methyl histidine, which has been used as an index of proteolysis in muscle. Both in rats given a free diet (17) and those given a 0.5% protein diet (18) the excretion of 3 methyl histidine was reduced. We conclude, then, that this slow loss of protein was primarily the result of a decrease in protein synthesis. The change in breakdown did not promote the change in protein mass but instead inhibited it.

## Effects of Starvation on Muscle Protein Breakdown

Measurement of the protein mass and the rate of protein synthesis in gastrocnemius muscle of 100 g rats during a two day fast gave results which were very similar to those for two days of a protein-free diet (16). There was no measurable fall in protein mass but there was an immediate fall in protein synthesis, corresponding to the cessation of normal growth. The picture changed, however, when starvation was studied for a longer period (19, Table 3). After two days the protein content of muscle started to fall at about 12% per day. Although the synthesis rate fell further between 2 and 4 days, the rapid loss of protein resulted mainly from an elevated rate of proteolysis. It has therefore been suggested the rate of protein breakdown rises as a response to serious stress, when protein is lost very rapidly; with more moderate forms of nutritional deprivation, such as the protein-free diet and starvation for two days, protein is only lost slowly and this is brought about by a decrease in synthesis (19).

These conclusions derived from indirect estimates of breakdown rate are consistent with measurements on incubated muscles *in vitro* (21). The rate of proteolysis, estimated from the release of amino acid into the incubation medium, remained the same when the donor animals were starved for one day, but increased when the donors were starved two days. The muscle protein mass in these experiments was also reduced at two days, probably because the rats used were smaller (75 g) than those in the experiments described above (100-120 g) which did not lose muscle protein till after 2 days. Hagen and Scow (22) demonstrated that the period of starvation before protein loss could be detected was longer with larger rats.

The results shown above are not, however, consistent with those recently derived from the urinary excretion of 3 methyl histidine (23). The rate of 3 methyl histidine excretion measured in the morning (after feeding) was higher than in the evening (before feeding). There were further increases after one and two days of fasting, so that the rate of excretion on the second day was 3 fold higher than the average during the day when food was given. The rats used in these experiments weighed more than 100 g and were therefore similar to those in Table 3, but the response to starvation for two days appeared to be very different. It is possible that the discrepancy relates to the particular muscles studied, since 3 methyl histidine excretion is an index of degradation of total body muscle protein. It is however worthwhile to examine in more detail the sources of error in the indirect method of determining breakdown from synthesis and growth, to see if these could be the cause of the discrepancy between the two studies.

## Sources of Error in the Indirect Method

It is to be expected that the result of an indirect calculation of breakdown from the difference between synthesis and growth will be particularly sensitive to errors arising from measurement of either synthesis or growth. The change in muscle protein is surprisingly difficult to measure accurately because of its slow rate. Measurements must therefore be made over a period of at least one day, and preferably over several days, if any confidence is to be placed in the result. With starvation it is difficult to decide when protein starts to be lost. Hagen and Scow (22) interpreted their data from starved rats as indicating no loss of protein for a period of time and a rapid loss thereafter, similar to the results in Table 3. By contrast Bird (24) fasted 200 g rats for 8 days and recorded a continuous but relatively slow loss in protein from the beginning. The best solution to this problem seems to lie in repeating the experiment more than once (see below).

A second source of error occurs because measurements of growth are made on a daily basis, but measurements of synthesis by constant infusion take only 6 h, which may not be representative of the whole

TABLE 4 The Effect of Starvation on Protein Turnover in Muscle: Measurement by Large Dose of [$^3$H]Valine

| Days of Starvation | Protein mass (% of control ± SEM) | $k_s$ (%/d) | $k_d$ (%/d) |
|---|---|---|---|
| 0 (day) | 100 ± 3 | 16.0 | 9.7 |
| 0 (night) | 98 ± 3 | 15.9 | - |
| 1 | 100 ± 4 | 11.1 | 11.1 |
| 2 | 97 ± 2 | 7.5 | 7.5 |
| 3 | 100 ± 4 | - | - |
| 4 | 91 ± 1 | 7.3 | 17.3 |

Rats weighing about 112 g were fed a diet containing 12% protein before starvation. Measurements of protein synthesis rate, by the method of Scornik (25) using [$^3$H]valine, were made in the morning, except the control group marked "night" which was fed *ad libitum* and measured at 22.00 h.

day. Estimates of synthesis during a single day demonstrated a statistically significant diurnal rythm (13). This rythm was, however, of very small amplitude, deviating by no more than 10% from the mean daily rate, and therefore it is unlikely that diurnal variations in synthesis would be a serious source of error in the determination of degradation rates.

Finally we must question the method used to measure synthesis. The main problem with the constant infusion method is the measurement of the specific activity of the free amino acids at the site of protein synthesis. For this purpose we have always used the specific activity of the free amino acid in the tissue homogenate, but this may be in error (for review see ref. 9). Consequently, to reduce this source of error, we again measured the effect of starvation on muscle protein and determined the rate of synthesis by giving the animals a very large dose of [$^3$H]valine which floods all the free amino acid pools so that they reach nearly the same specific radioactivity (25)

The protocol chosen was that described by Scornik (25). He injected mice with [$^{14}$C]leucine in doses varying from 33 to 245 μmoles/100 g body wt, while maintaining the specific activity constant. Livers were taken after 5 mins and the incorporation of isotope into protein determined. A graph of the reciprocal of dose against the reciprocal of incorporation gave a straight line whose intercept was the incorporation at infinite dose ( i.e. when 1/ dose = 0, dose = ∞). The rate of protein synthesis was then calculated from this intercept value for incorporation on the assumption that at infinite dose the specific activity at the site of protein synthesis would be equal to that of the injected amino acid. This method, therefore, avoids the problem of measuring the precursor specific activity.

The results obtained by the above procedure using [$^3$H]valine to measure protein synthesis in rats starved for 4 d are shown in Table 4. The protein mass in gastrocnemius muscle did not change measurably until after the 3rd day, then there was a fall of 9% between the 3rd and 4th days. The delay before protein was lost therefore seems to be longer than in the experiment shown in Table 3, perhaps because of the lower protein content of the diet prior to starvation. The rate of synthesis in fed rats was the same whether measurements were made during the day or during the night. On the 4th day of starvation, when protein started to be lost rapidly, the calculated rate of protein breakdown increased. Therefore, the conclusions reached in the two experiments shown in Tables 3 and 4, using two different methods for measuring the rate of protein synthesis, are basically the same. This increases our confidence that the indirect approach for estimating protein breakdown in muscle from measurements of synthesis and growth of protein mass is a valid and useful method.

## REFERENCES


1 D.J. Millward, Protein turnover in skeletal muscle. I. Clin. Sci. 39, 577 (1970).

2 R.W. Swick & M.M. Ip, Measurement of protein turnover in rat liver with $^{14}$C carbonate. J. Biol. Chem. 249, 6836 (1974).

3 R.T. Schimke, The importance of both synthesis and degradation in the control of arginase levels in the rat liver. J. Biol. Chem. 239, 3808 (1964).

4 R. Druyan, B. DeBernard & M. Rabinowitz, Turnover of cytochromes labelled with δ-aminolevulinic acid-$^3$H in rat liver. J. Biol. Chem. 244, 5874 (1969).

5 P.J. Garlick, J.C. Waterlow & R.W. Swick, Measurement of protein turnover in rat liver. Biochem. J. 156, 657 (1976).

6 R.T. Schimke, (1970) Regulation of protein degradation in mammalian tissues. In: Mammalian Protein Metabolism Vol IV. Ed. H.N. Munro. Ch. 32. p 178 Academic Press, New York.

7 R.T. Schimke & D. Doyle, Control of enzyme levels in animal tissues, Ann. Rev. Biochem. 39, 929 (1970).

8 O.A. Scornik & V. Botbol, Role of changes in protein degradation in the growth of regenerating livers. J. Biol. Chem. 251, 2891 (1976).

9 J.C. Waterlow, P.J. Garlick & D.J. Millward (1978) Protein Turnover in Mammalian Tissues and in the Whole Body. Elsevier, Amsterdam.

10 R.D. Conde & O.A. Scornik, Role of protein degradation in the growth of livers after a nutritional shift. Biochem. J. 158, 385 (1976).

11 J.C. Waterlow & J.M.L. Stephen, The effect of low protein diets on the turnover rates of serum, liver and muscle proteins in

the rat, measured by continuous infusion of L[$^{14}$C]lysine. Clin. Sci. 35, 287 (1968).

12 P.J. Garlick & I. Marshall, A method for measuring brain protein synthesis. J. Neurochem. 19, 577 (1972).

13 P.J. Garlick, D.J. Millward & W.P.T. James, The diurnal response of muscle and liver protein synthesis *in vivo* in meal-fed rats. Biochem. J. 136, 935 (1973).

14 T. Peters, Jr. & J.C. Peters, The biosynthesis of rat serum albumin. VI. J. Biol. Chem. 247, 3858 (1972).

15 M. Haider & H. Tarver, Effect of diet on protein synthesis and nucleic acid levels in rat liver. J. Nutr. 99, 433 (1969).

16 P.J. Garlick, D.J. Millward, W.P.T. James & J.C. Waterlow, The effect of protein deprivation and starvation on the rate of protein synthesis in tissues of the rat. Biochim. Biophys. Acta 44, 71 (1975).

17 R. Funabiki, Y. Watanabe, N. Nishizawa & S. Hareyama, Quantitative aspects of the myofibrillar protein turnover in transient state on dietary protein depletion and repletion revealed by urinary excretion of N-methyl histidine. Biochim. Biophys. Acta 451, 143 (1976).

18 L.N. Haverberg, L. Deckelbaum, C. Bilmazes, H.N. Munro & V.R. Young, Myofibrillar protein turnover and urinary N-methyl histidine output. Biochem. J. 152, 503 (1975).

19 D.J. Millward, P.J. Garlick, D.O. Nnanyelugo & J.C. Waterlow, The relative importance of muscle synthesis and breakdown in the regulation of muscle mass. Biochem. J. 156, 185 (1976).

20 J.C. Waterlow, P.J. Garlick & D.J. Millward (1977) Amino acid supply and Protein turnover. In: Clinical Nutrition Uptake - Amino Acids. Eds. H.L. Greene, M.A. Holliday & H.N. Munro. p. 1 American Medical Association. Chicago.

21 J.B. Li & A.L. Goldberg, Effects of food deprivation on protein synthesis and degradation in rat skeletal muscle. Am. J. Physiol. 231, 441 (1976).

22 S.N. Hagan & R.O. Scow, Effect of fasting on muscle proteins and fat in young rats of different ages. Am. J. Physiol. 188, 91 (1957).

23 S.J. Wassner, S. Orloff & M.A. Holliday, Protein degradation in muscle: response to feeding and fasting in growing rats. Am. J. Physiol. 233, E119 (1977)

24 J.W.C. Bird (1975) Skeletal Muscle Lysosomes. In: Lysosomes in Biology and Pathology. Vol. IV. Eds. J. Dingle & R. Dean. Ch. 4. North Holland Pub. Co. Amsterdam.

25 O.A. Scornik, *In vivo* rate of translation by ribosomes of normal and regenerating liver. J. Biol. Chem. 249, 3876 (1974).

ACKNOWLEGEMENTS

The authors are grateful to the Royal Society and to the Wellcome Trust for financial support.

# SELECTIVE CONTROL OF PROTEINASE ACTION IN YEAST CELLS

Helmut Holzer
Institut für Toxikologie und Biochemie der Gesellschaft für Strahlen- und Umweltforschung München und Biochemisches Institut der Universität Freiburg; D-7800 Freiburg im Breisgau, Germany

## ABSTRACT

The proteolytic system in yeast, consisting of at least six proteinases and three groups of inhibitors specific for the proteinases A and B, and carboxypeptidase Y, respectively, is described. Participation of proteinases and inhibitors in the following processes is discussed: Activation of precursors of chitinsynthase and carboxypeptidase Y by limited proteolysis; catabolite inactivation of selected enzymes after addition of glucose to acetate or galactose grown cells; catabolite modification of the galactose and maltose uptake system; degradation of certain enzymes at the transition from exponential to stationary growth and to spore formation, respectively.

The intracellular proteinases from *Saccharomyces cerevisiae* known at present are summarized in Table 1. The proteinases A, B and carboxypeptidase Y are localized in the yeast-vacuoles. Intracellular localization of the other proteinases is not yet completely clarified. The present knowledge on the biological function of the endogeneous yeast-proteinases may be summarized as follows:

## I. Complete Degradation of Proteins

General protein turnover; catabolite inactivation of cytoplasmic malate dehydrogenase, fructose 1,6-bisphosphatase and phosphoenol pyruvate carboxykinase; degradation of NADP-glutamate dehydrogenase at release from glucose protection in candida yeast; degradation of certain enzymes at the transition from exponential to stationary phase; degradation of NAD-glutamate dehydrogenase and some other enzymes at transition from vegetative growth to spore formation.

## II. Limited Proteolysis

Prochitinsynthase → chitinsynthase; procarboxypeptidase Y → carboxypeptidase Y; catabolite modification of the galactose and maltose uptake system; formation of subunits of cytochrome-oxidase from precursor protein.

TABLE 1 Intracellular Proteolytic Enzymes from *Saccharomyces cerevisiae*

| Enzyme | Characteristics |
|---|---|
| Proteinase A | Acid endo-peptidase<br>MW = 30-40 000 |
| Proteinase B | Serine endo-peptidase<br>MW = 32-44 000 |
| Carboxypeptidase Y | Serine exo-peptidase<br>MW = 61 000 |
| Carboxypeptidase S | Exo-peptidase ($Me^{++}$) |
| Aminopeptidase I | Exo-peptidase ($Zn^{++}$)<br>MW = 200 000 |
| Aminopeptidase II | Exo-peptidase ($Me^{++}$)<br>MW = 34 000 |

TABLE 2 Proteinase Inhibiting Polypeptides from Yeast

| Inhibitor | Isoelectric point | Specific inhibition of | Hydrolyzed by | Molecular weight | Stability |
|---|---|---|---|---|---|
| $I^A2$ | 5.7 | Proteinase A | Proteinase B | 7,700 | Heat- and acid-resistant |
| $I^A3$ | 6.3 | | | | |
| $I^B1$ | 8.0 | Proteinase B | Proteinase A | 8,500 | Heat- and acid-resistant |
| $I^B2$ | 7.0 | | | 8,500 | |
| $I^B3$ | 4.6 | | | 14,000 | |
| $I^C$ | 6.6 | Carboxypeptidase Y | Proteinases A and B | 23,800 | Heat- and acid-labile |

Specific endogeneous inhibitors of the proteinases A and B, and carboxypeptidase Y have been isolated and characterized in the last 4 years (cf. Table 2). The inhibitors are polypeptides, they contain no carbohydrate or other nonpeptide material. The iso-inhibitor $I^B3$ has been discovered very recently (1), the other inhibitors have been reviewed by Wolf and Holzer (2). Proteinases A and B, and carboxypeptidase Y are found in the vacuoles, whereas the inhibitors are localized in the cytosol. The high proteinase specificity of the proteinase inhibitors suggests a specific function of these inhibitors in the control of intracellular proteinase activity. At present the following functions of the inhibitors are considered:

1. Protection of extravacuolar space against unwanted proteolysis by leaky or lysed vacuoles (containing the endoproteinases A and B, and carboxypeptidase Y),

2. control of proteinase B which converts (in vitro) by limited proteolysis prochitinsynthase to chitinsynthase (3,4), and procarboxypeptidase Y to carboxypeptidase Y (5).

A cascade model for the interaction of proteinases and inhibitors at activation of prochitinsynthase which is firmly bound to the plasma membrane is shown in Fig. 1. The single steps of the cascade,

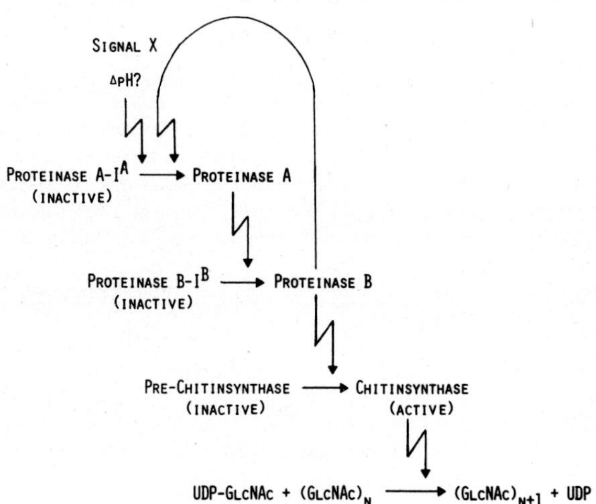


Fig. 1. Postulated cascade for activation of chitin synthesis

such as dissociation of $I^A$-inhibited proteinase A by a decrease in pH, hydrolysis of the A-inhibitor by proteinase B, activation of $I^B$-inhibited proteinase B by proteinase A, which hydrolyses $I^B$, and the activation of prochitinsynthase by proteinase B catalyzed limited proteolysis have been demonstrated in vitro. The cascade-ordered cooperation of the proteinases and inhibitors, shown in Fig. 1 is, however, hypothetical.

After addition of glucose to yeast cells grown on acetate, galactose or another non-glucose related substrate not only the well known catabolite repression of enzymes of gluconeogenesis (6), but also a rapid inactivation and modification of certain enzymes is observed (7). The enzymes known at present to be inactivated or irreversibly modified are listed in Table 3.

TABLE 3 Glucose-Effected Inactivation and Irreversible Modification of Enzymes

I CATABOLITE INACTIVATION

CYTOPLASMIC MALATE DEHYDROGENASE

FRUCTOSE 1,6-BISPHOSPHATASE

PHOSPHOENOLPYRUVATE CARBOXYKINASE

II CATABOLITE MODIFICATION

GALACTOSE UPTAKE SYSTEM

α-GLUCOSIDE (MALTOSE) PERMEASE

During catabolite inactivation the catalytic activity of the respective enzymes disappears completely; during catabolite modification only the affinity of the two uptake systems for the respective sugars is decreased, whereas $V_{max}$ is not changed (8,9). The data for the galactose uptake system are summarized in Table 4. As

TABLE 4 Galactose Uptake System of Saccharomyces cerevisiae

| | $V_{MAX}$ (μMOL GAL X $MIN^{-1}$ X G WET $WEIGHT^{-1}$) | $K_M$ (mM GAL) |
|---|---|---|
| GALACTOSE-GROWN | 3.9 | 3.6 |
| 4 H WITH GLUCOSE | 3.9 | 11 |

shown in experiments with cycloheximide and other inhibitors of protein synthesis, a reversal of both, catabolite inactivation and catabolite modification after washing out glucose and resuspension in a glucose-free growth medium depends on de novo protein synthesis. This finding suggests a common mechanism, namely proteolysis, to be responsible for catabolite inactivation as well as catabolite modification. For cytoplasmic malate dehydrogenase the parallelism of disappearance of catalytic activity and reactivity with antibodies, demonstrated by Neeff et al. (10) further supports the idea that proteolysis is the mechanism of inactivation. The modification of the two permeases (see Table 3) could well be understood as a consequence of limited proteolysis. A necessity of de novo protein synthesis for reversal of the modification would then, however, be difficult to understand because at present no such case is known.

Matern and Holzer (9) therefore discuss the possibility that analogous to the functioning of lactalbumin as a "modifying protein" in the lactose-synthesizing system (11) a non-catalytic protein which modifies the affinity of the permeases for galactose and maltose, respectively, is subject to catabolite inactivation. Catabolite modification of the permeases would then be the consequence of catabolite inactivation of modifier proteins. A schematic presentation of this hypothesis is shown in Fig. 2. How the enzymes (and perhaps

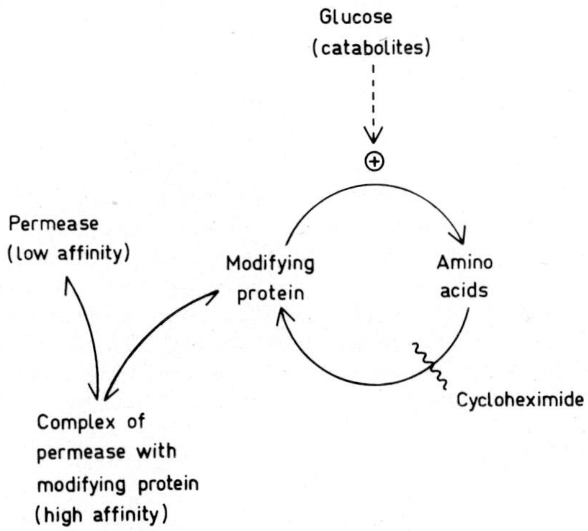


Fig. 2. Hypothesis for modification of a permease by catabolite inactivation of a modifying protein

the modifying proteins) which are proteolyzed in catabolite inactivation are selected from the thousands of enzymes available in the cell is at present completely unknown. Even the fundamental question if proteolysis takes place in the vacuoles after uptake of selected enzymes or if the proteinases leave the vacuoles and act selectively on the respective enzymes in the extravacuolar space, cannot be answered at present.

Transition from vegetative growth to spore formation is another situation in which selective inactivation of certain enzymes takes place. Results from Betz and Weiser (12) are listed in Table 5.

TABLE 5 Specific Activity of some Yeast Enzymes after Suspension in Sporulation Medium

| | 0 h | 24 h |
|---|---|---|
| Malate dehydrogenase | 2.6 | 1.0 |
| Isocitrate lyase | 0.176 | 0.035 |
| NAD-dependent glutamate dehydrogenase | 0.36 | 0.02 |
| Fructose bisphosphatase | 0.127 | 0.030 |
| Alcohol dehydrogenase | 3.06 | 1.07 |
| Hexokinase | 1.77 | 0.76 |
| Glutamate oxalacetate transaminase | 0.295 | 0.125 |

Inactivation of the enzymes shown in Table 5 is prevented by cycloheximide and carbonylcyanide chlorophenylhydrazone, substances known to inhibit protein degradation in sporulating yeast cells (13). The mechanism of inactivation is therefore very probably proteolysis. This idea is supported by the finding that specific activity of the proteinases A and B increases drastically after transfer of vegetative growing cells to sporulation medium (13). Again nothing is known about the mechanisms of selection of enzymes for proteolysis.

## REFERENCES

(1) H. Matern, U. Weiser and H. Holzer, $I^B3$: A new inhibitor of proteinase B from *Saccharomyces carlsbergensis*, 12th FEBS-Meeting Dresden, DDR, 1978, abstract accepted.

(2) D.H. Wolf and H. Holzer, Proteolysis in yeast, in: *Transport & Utilization of Amino Acids, Peptides & Proteins by Microorganisms* (J.W.Payne, ed.) Wiley, Chichester-England, 1978.

(3) R.E. Ulane and E. Cabib, The activating system of chitin synthetase from *Saccharomyces cerevisiae*, *J. Biol. Chem.* 251, 3367 (1976).

(4) A. Hasilik and H. Holzer, Participation of the tryptophan synthase inactivating system from yeast in the activation of chitin synthase, *Biochem. Biophys. Res. Commun.* 53, 552 (1973).

(5) A. Hasilik and W. Tanner, Biosynthesis of the vacuolar yeast glycoprotein carboxypeptidase Y. Conversion of precursor into enzyme, Eur. J. Biochem. 85, 599 (1978).

(6) B. Magasanik, Catabolite repression, Cold Spring Harbor Symp. Quant. Biol. 26, 249 (1961).

(7) H. Holzer, Catabolite inactivation in yeast, Trends in Biochemical Sciences (TIBS) 1, 178 (1976).

(8) C.P.M. Görts, Effect of glucose on the activity and the kinetics of the maltose uptake system and of α-glucosidase in Saccharomyces cerevisiae, Biochim. Biophys. Acta 184, 299 (1969).

(9) H. Matern and H. Holzer, Catabolite inactivation of the galactose uptake system in yeast, J. Biol. Chem. 252, 6399 (1977).

(10) J. Neeff, E. Hägele, J. Nauhaus, U. Heer and D. Mecke, Application of an immunoassay to the study of yeast malate dehydrogenase inactivation, Biochem. Biophys. Res. Commun., 80, 276 (1978).

(11) R.L. Hill, R. Barker, K.W. Olsen, J.H. Shaper and I.P. Trayer, Lactose synthetase: structure and function, in: Metabolic Interconversion of Enzymes (O.Wieland, E.Helmreich, H.Holzer, eds.) Springer Berlin.Heidelberg.New York 1972, p.331.

(12) H. Betz and U. Weiser, Protein degradation during yeast sporulation. Enzyme and cytochrome patterns, Eur. J. Biochem. 70, 385 (1976).

(13) H. Betz and U. Weiser, Protein degradation and proteinases during yeast sporulation, Eur. J. Biochem. 62, 65 (1976).

# STUDIES OF THE MECHANISM AND SELECTIVITY OF INTRACELLULAR PROTEIN BREAKDOWN

Alfred L. Goldberg, Yoel Klemes, Koko Murakami,
Nicola Neff and Richard Voellmy
Department of Physiology, Harvard Medical School
Boston, Massachusetts 02115

The rate of degradation of cell proteins is determined both by structural features of the individual molecules and by the overall nutritional or endocrine status of the cell (1,2). Thus bacteria and mammalian cells degrade proteins with highly abnormal structures very rapidly. Such proteins may result from mutations, biosynthetic errors, incorporation of amino acid analogs, etc. The breakdown of abnormal proteins is rapid whether the cells are growing or starving. By contrast, the degradation of normal proteins in Escherichia coli increases 2-3 fold upon deprivation for amino acids, aminoacyl tRNA, or a carbon source (2-5). Experiments with $rel^+/rel^-$ and $spoT^+/spoT^-$ strains indicate that this acceleration of proteolysis requires the accumulation of guanosine tetraphosphate within the bacteria (4,5,7). This nucleotide is known to inhibit synthesis of ribosomal RNA, tRNA, and lipids and to regulate a variety of other growth-related processes.

Guanosine tetraphosphate is synthesized at increased rates when the ribosome lacks a sufficient supply of aminoacyl tRNA for continued protein synthesis (6). ppGpp turns over in $spoT^+$ cells with a half-life of about 30 seconds. If production of guanosine tetraphosphate in amino acid-deprived cells is prevented in $spoT^+$ cells with tetracycline, then protein breakdown in the starving cell fall to basal levels as rapidly as can be measured (in less than 1-2 minutes) (7). The rapidity of these changes in protein degradation indicates that guanosine tetraphosphate must act by altering the activity of the cell's proteolytic enzymes or alternatively by affecting the sensitivity of cell proteins to this nucleotide (7).

The rapid degradation of analog-containing proteins has been studied in both E. coli and rabbit reticulocytes (1,2,8,13). Although hemoglobin normally is an unusually stable protein, incorporation of the valine-analog, α-aminochlorobutyric acid, by reticulocytes leads to the production of a globin of normal size that is degraded with a half-life of 10-15 min. (12,13). Such analog-containing globins aggregate into large molecular complexes prior to their rapid degradation. On gel filtration, such complexes show an apparent (10) molecular weight of 1 to 2 million. In reticulocytes, these structures appear to be smaller than the analgous dense granules containing abnormal proteins that we described earlier in bacteria (8). However in both eukaryotic and procaryotic cells, the dissapearance of such aggregated material can account for all the proteins being degraded to amino acids.

Studies with inhibitors of energy metabolism indicate that the subsequent hydrolysis of the abnormal protein requires high energy phosphates(2,14).To clarify the actual pathway of proteolysis and to elucidate on the nature of this apparent energy requirement, we developed cell-free conditions for the selective breakdown of abnormal proteins in reticulocyte (12) and bacterial lysates (16). In reticulocyte extracts, the degradation of analog-containing proteins to free amino acids is stimulated by ATP, which may account for the energy requirement of proteolysis in vivo. The responsible proteolytic activity differs in many respects for the lysosomal proteases (12). It shows an optimal pH at 8, is found in the 100,000g supernatant, and is not sensitive to agents that inhibit lysosomal proteases such as leupeptin and chymostatin. By contrast, these inhibitors were found to reduce overall protein breakdown in skeletal muscle (15) or hepatocytes without affecting the degradation of abnormal proteins or of normal proteins with short half-lives (16). These agents also could inhibit the degradation of extracellular proteins, such as asialofetuin. (This protein is taken up by receptor-mediated endocylosis, and its subsequent degradation occurs within the lysosome.) Thus eukaryotic cells appear to contain multiple proteolytic systems that serve distinct degradative functions.

We have also demonstrated an analogous ATP-dependent degradative system in E. coli extracts prepared by gentle lysis techniques (16, 17). The responsible activity responds to ATP and other nucleotides by increasing the degradation of globin or nonsense fragments of β-galactosidase several fold. The responsible ATP-dependent proteolytic activity appears to be particulate and associated with the cell membrane (17). By gel filtration, this proteolytic activity can be separated from soluble peptidases. The membrane associated enzyme degrades proteins to large peptides (greater than 1500MW) which are then degraded by soluble enzymes to free amino acids. A number of mutations are known which can eliminate these peptidase reactions without affecting the ATP-dependent degradation of proteins. Together these experiments indicate that intracellular proteolysis is a multi-step process in which an initial ATP-dependent endoproteolytic cleavage appears to be the rate-limiting event. This conclusion is in accord with earlier suggestions based on studies of a specific abnormal protein in intact cells (18).

## REFERENCES

1. Goldberg, A.L. and J.F. Dice. Intracellular protein degradation in mammalian and bacterial cells. Ann. Rev. Biochem. 43: 835-869 1974.
2. Goldberg, A.L. and A.C. St. John. Intracellular protein degradation in mammalian and bacterial cells: Part II. Ann. Rev. Biochem. 45: 747-803, 1976.
3. Mandelstam, J. The Intracellular turnover of protein and nucleic acids and its role in biochemical differentiation. Bacteriol. Rev. 24: 289-308, 1960.
4. St. John, A.C., A.L. Goldberg, K. Conklin, and E. Rosenthal. Further evidence for the involvement of charged tRNA and guanosine tetraphosphate in the control of protein degradation in E. coli. J. Biol. Chem. 253: 3945-3951, 1978.
5. St. John, A.C. and A.L. Goldberg. Effects of reduced energy production on protein degradation, Guanosine tetraphosphate, and RNA synthesis in E. coli. J. Biol. Chem. 253: 2705-2711, 1978.
6. Cashel, M. Regulation of bacterial ppGpp and pppGpp. Ann. Rev. Microbio. 29: 301-318, 1975.
7. Voellmy, R., and A.L. Goldberg. Membrane-association of the ATP-dependent proteolytic activity in E. coli. Submitted for publication, 1978.
8. Prouty, W.F., Karnovsky, M.J. and Goldberg, A.L. Degradation of abnormal proteins in E. coli: Formation of protein inclusions in cells exposed to amino acid analogs. J. Biol. Chem. 250: 1112-1122, 1975.
9. Goldberg, A.L., J.D. Kowit, J.K. Etlinger, and Y. Klemes. Selective degradation of abnormal proteins in animal and bacterial cells. In "Protein Turnover and Lysosomal Function", SUNY Symposium, Doyle, D., Segal, H., Academic Press, (In press) 1978.
10. Klemes, Y, and A.L. Goldberg. Physical and chemical properties of proteins rapidly degraded in rabbit reticulocytes. Submitted for publication, 1978.
11. Neff, N., and A.L. Goldberg. Effects of protease inhibitors on the degradation of different classes of proteins in cultured rat hepatocytes. Submitted for publication, 1978.
12. Etlinger, J. and A.L. Goldberg. A soluble, ATP-dependent proteolytic system responsible for the degradation of abnormal proteins in reticulocytes. Proc. Nat. Acad. Sci. 74: 54-58, 1977.
13. Rabinovitz, M. and J.M. Fisher. Characteristics of the inhibition of hemoglobin synthesis in rabbit reticulocytes by threo-α-amino-β-chlorobutyric acid. Biochim. Biophys. Acta, 91: 313-322, 1964.
14. Olden, K., and Goldberg, A.L. Studies of the nature of the energy requirement for protein degradation in bacteria. Biochem. Biophys. Acta. (In press) 1978.

15. Libby, P. and Goldberg, A.L. Leupeptin, a protease inhibitor, decreases protein degradation in normal and diseased muscles. Science, 199: 534-536, 1978.
16. Murakami, K., R. Voellmy, and A.L. Goldberg. An ATP-stimulated proteolytic system that degraded abnormal proteins in *E. coli* extracts. Submitted for publication, 1978.
17. Voellmy, R., and A.L. Goldberg. Membrane-association of the ATP-dependent proteolytic activity in *E. coli*. Submitted for publication, 1978.
18. Kowit, J.D. and A.L. Goldberg. Intermediate steps in the degradation of a specific abnormal polypeptide in *E. coli*. J. Biol. Chem. 252: 8350-8357, 1977.

## ACKNOWLEDGEMENT

These studies have been supported by grants to Alfred L. Goldberg from the National Institute of Neurological Disease and Stroke, the Muscular Dystrophy Associations of America, and the National Aeronautics and Space Administration. During the course of these experiments, Dr. Voellmy held a fellowship from the Swiss National Fund, Dr. Murakami from the Massachusetts Heart Assoc., and Dr. Nicola Neff from the Muscular Dystrophy Association of America. The authors are grateful to Julia Smith for her assistance in preparing this manuscript.

# THE ROLE OF CATHEPSINS L AND H FROM RAT LIVER LYSOSOMES IN PROTEIN DEGRADATION

H. Kirschke, J. Langner, B. Wiederanders,
S. Ansorge, P. Bohley
Physiologisch-chemisches Institut der
Martin-Luther-Universität Halle-Wittenberg,
Halle(Saale), GDR

Intracellular protein degradation is necessary for changes in the composition of enzymes in all cells, otherwise the protein concentration in the cells will rise uncontrollably. Only in rapidly growing cells is the continual degradation of proteins less necessary, since the cells can outgrow their enzyme complement by diluting the preexisting enzymes. In animal tissues which are not normally growing, however, degradation of proteins is the main mechanism available for their removal. Protein turnover has been demonstrated in all cells and all cell organelles so far investigated. This protein degradation occurs largely intracellularly and is selective. High turnover rates have been found for enzymes from the cytosol. On the other hand the autoproteolysis of cytosol is extremely weak in comparison to that of the other cell organelles (Ref. 1). The very high turnover rates in vivo of some of the cytosol proteins cannot be explained in terms of their degradation by cytosol proteinases alone. Therefore an intracellular cooperation in vivo was assumed, in which cytosol proteins are substrates for protein breakdown systems in other organelles.

In-vitro investigations of these postulations require conditions which closely simulate those in vivo, otherwise the interpretation of such experiments in regard to in-vivo protein turnover cannot be conclusive. For the investigation of intracellular protein breakdown, therefore, we used intracellularly occurring cytosol proteins as substrates, incubated at neutral pH values in the presence of naturally occurring ions and effectors (e.g. phosphate buffer and glutathione). Of course some cathepsins (e.g. cathepsin L) have highly acidic pH optima, but such pH values have not been detected in hepatocytes. Even in lysosomes the mean pH was 6,2 (Ref. 2). However, the in-vitro simulation of in-vivo conditions is necessarily inadequate with respect to the rather considerable natural membrane barieres. But the following calculation may possibly permit a rough comparison of results from proteolysis measured in vitro with the known turnover rates in vivo. From the release of trichloroacetic acid-soluble nitrogen we can calculate the percentage of protein degraded. If we assume a weight ratio between cytosol proteins and lysosomal proteins of 40:1 for rat liver, then the activities determined in our laboratory for lysosomes alone might be almost sufficient for the degradation of more than 5% of cytosol proteins per hour. This value approaches the normal rate of degradation in vivo. Moreover, assuming for rat liver a ratio between cytosol proteins and microsomal proteins of 3:1, the activities detected in microsomes could just possibly account for the degradation of about 5% of cytosol proteins per hour. But we have to take into consideration that, in contrast to the lysosomes, the microsomal fraction is less active in the hydrolysis of short-lived cytosol proteins (Ref. 3). Therefore it remains to be shown whether lysosomes or microsomes degrade the main part of the cytosol proteins in vivo. We decided first to investigate proteolytic enzymes in lysosomes using cytosol proteins as substrate.

From similar calculations it follows clearly that nuclei, mitochondria and cytosol cannot contribute greatly to the general protein breakdown. The proteinases within these cell organelles seem to be more specific and hydrolyze only special proteins, for instance, the group-specific proteinases of mitochondria (however, which do not inactivate all pyridoxine enzymes) (Ref. 4), the insulin-glucagon-"specific" enzyme of cytosol (Ref. 5) and a proteinase of ribosomes (Ref. 6).

Now we will summarize the properties of those lysosomal proteinases which are active at pH 6. This pH value has been established in hepatocyte lysosomes.
A cell fraction 10 fold enriched in lysosomes has been obtained from rat liver homogenate by a special fractionation scheme (Ref. 7). Water treatment of this fraction and high-speed centrifugation resulted in an extract 50-100 fold enriched in solubilized lysosomal enzymes. By fractionation on Sephadex G-75 the bulk of soluble lysosomal proteins proved to be of high molecular weight (>80,000). Only 15% of the proteins in the extract had molecular weights of 15-30,000. The distribution pattern of proteolytic activity in splitting cytosol proteins at pH 6 is quite different. The high molecular weight proteins accounted for only 10-12% of total proteolytic activity and the proteins of molecular weight

15-30,000 accounted for 75-80% of the activity (Ref. 8). Cathepsins E and D are eluted corresponding to their molecular weights and their part in proteolytic activity at pH 6 is only small. The high activity is due to cathepsins L, B and H which all have molecular weights in the range of 20-30,000.

The properties of cathepsin B are wellknown from studies by Otto (Ref. 9) and by Barrett (Ref. 10). We have recently described the isolation and properties of the cathepsins L and H (Ref. 11, 12). Cathepsins L, B and H are all thiolproteinases. Cathepsin H is a glycoprotein and the last step in the isolation procedure is its desorption from Concanavalin-A-Sepharose. The isoelectric point is at pH 7,1 and above this pH cathepsin H is very unstable. The same behaviour holds true for cathepsins L and B. The pH activity optimum noted here has been measured with the substrates given in Table 1.

TABLE 1 Properties of Main Lysosomal Cathepsins
Activity at pH 6

| | Cathepsin L | Cathepsin B | Cathepsin H |
|---|---|---|---|
| Mol. wt ($x10^{-3}$) | 24 | 28 | 28 |
| $p_I$ | 5,8-6,1 | 5,4-5,6 | 7,1 |
| pH-Optimum | 5,5 | 5,5 | 6,0 |
| Yield from 100 rat livers (mg) | 2 | 7 | 6 |
| Main substrates | proteins | Bz-Arg-2-NNap<br>proteins | Bz-Arg-2-NNap<br>Arg-2-NNap<br>proteins |
| Leupeptin $K_i$ (M) | $2 \times 10^{-8}$ | $2 \times 10^{-8}$ | $5 \times 10^{-6}$ |

The three enzymes differ from one another precisely in their specificity of splitting substrates. Cathepsin H is an endopeptidase which hydrolyzes proteins and synthetic low molecular weight substrates (e.g. Bz-Arg-2-NNap, esters and amides). But this enzyme also hydrolyzes aminopeptidase substrates, e.g. Arg-2-NNap. Therefore we have designated this enzyme an endoamino-peptidase, thus establishing a new group of proteolytic enzymes. Cathepsin B splits proteins and the wellknown synthetic substrates (e.g. Bz-Arg-2-NNap, esters and amides). Moreover, it should be mentioned that cathepsin B as well as cathepsin L might also be involved in the conversion of proteinogens into active proteins as was shown in our laboratory by the conversion of proinsulin to insulin or des-Ala-insulin (Ref. 13). Leupeptin and also chymostatin, are powerful inhibitors of cathepsins L and B and do not inhibit cathepsin H so strongly. In contrast to the other two cathepsins, cathepsin L hydrolyzes proteins in particular and nearly no synthetic substrates. [$^{3}$H]-labelled cytosol proteins, azocasein and [$^{125}$I]-labelled glucagon were the main substrate proteins used in our experiments. Certainly, cathepsin L has a greater importance in protein degradation than cathepsin B and cathepsin H: its specific activity in splitting proteins is 10–15 times higher than those of cathepsin B and cathepsin H.

We have calculated the share which of each the proteinases in the total activity of the lysosomal extract using cytosol proteins as substrates. The results are given in Fig. 1.

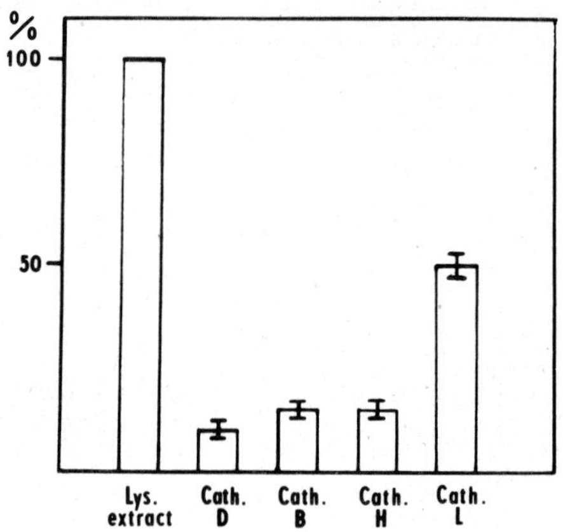


Fig. 1. Participation of lysosomal enzymes in the degradation of cytosol proteins at pH 6,1.

The share of the cathepsins in degrading cytosol proteins is expressed as a percentage of the total activity observed in the lysosomal extract. It has been calculated after enzyme isolation on the basis of their specific activities and their amounts within the lysosomal extract.

Cathepsin L has by far the highest share in the total activity of the soluble part of lysosomes at pH 6, whereas cathepsins B and H each contribute only 15–20% of the total activity in degrading cytosol proteins. Cathepsin D has only a weak activity at this pH. Furthermore, these enzymes degrade short-lived and long-lived cytosol proteins at different rates, as demonstrated in Fig. 2.

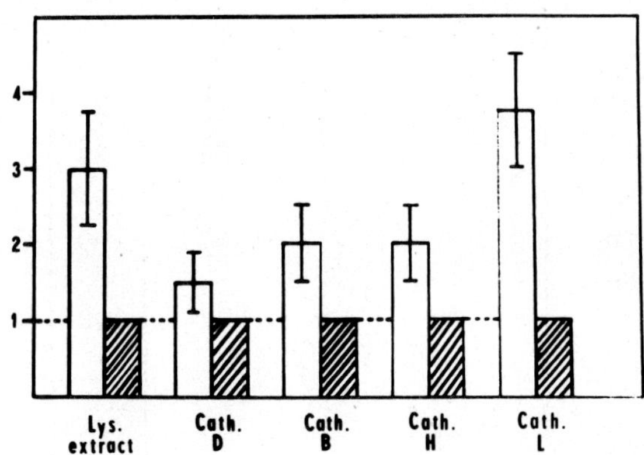


Fig. 2. Ratio of degradation in vitro of short-lived cytosol proteins to long-lived ones by lysosomal enzymes.

The value 1.0 is assigned to the degradation of long-lived cytosol proteins, labelled in vivo with [$^{14}$C]-guanidino-Arg ▨. Short-lived cytosol proteins were labelled with [$^{3}$H]-amino acids and their degradation is given as a ratio of the degradation of long-lived ones ☐.

Strikingly cathepsin L prefers to hydrolyze short-lived proteins. But this is not direct evidence of an important role played by cathepsin L in the degradation of short-lived cytosol proteins because other proteinases which do not occur within cells, such as trypsin and chymotrypsin, also prefer to hydrolyze short-lived proteins rather than long-lived ones. The explanation may reside in the specificity of the enzymes and in certain common properties of these substrate proteins. This point will be discussed later.

First, we shall give the results of our studies using proteinase inhibitors. These studies have been carried out to obtain more evidence for the involvement of the cathepsins L, B and H in overall proteolysis (Table 2). Most of the inhibitors were kindly provided by Prof. Umezawa and we wish to thank him very much.

TABLE 2 The Degradation of Cytosol Proteins in the Presence of Inhibitors

| Addition | Homogenate | Lys. extract | Cathepsins H | B | L |
|---|---|---|---|---|---|
| none | 100 | 100 | 100 | 100 | 100 |
| 1. Leupeptin | 52 | 30 | 17 | 35 | 4 |
| 2. Chymostatin | 39 | 27 | 44 | 8 | 2 |
| 3. Elastatinal | 68 | 86 | 34 | 17 | 92 |
| 4. Antipain | 58 | 31 | 19 | 17 | 4 |
| 5. Pepstatin | 77 | 97 | 97 | 112 | 105 |
| 6. Soya-bean trypsin inhibitor | 97 | 101 | 100 | 100 | 140 |
| 7. Phenyl-methane-sulphonyl fluoride | 95 | 93 | 60 | 80 | 100 |
| 8. 1.+2.+5. | 20 | 10 | 14 | 5 | 1 |
| 9. 1.+2.+3.+5. | 18 | 10 | 14 | 0 | 0 |

The residual activity is expressed as a percentage of that observed in control incubations without inhibitors. Homogenate of liver cells, lysosomal extract and the isolated proteinases were incubated with the inhibitors and [$^3$H]-labelled cytosol proteins as substrate. Each inhibitor was present at $10^{-4}$ M.

The following results are note-worthy:
The most powerful inhibitors proved to be leupeptin, chymostatin and antipain. The inhibition of the proteolytic activity of the homogenate by elastatinal seems to indicate an enzyme or a group of enzymes not of lysosomal origin. The weak inhibition by pepstatin, an inhibitor of carboxylproteinases, confirms the small degree of involvement of cathepsin D in overall proteolysis at pH 6. Serine proteinase inhibitors, such as Soya-bean trypsin inhibitor and phenyl-methane-sulphonyl fluoride, have little or no inhibitory effect on proteolysis at pH 6. Combinations of the inhibitors almost completely suppress the protein degradation in the homogenate as well as in the lysosomal extract.

The same results have been established using other substrates, such as azocasein or [$^{125}$I]-casein coupled to Sepharose, a useful substrate which allows the detection of large degradation products. The results of similar inhibitory studies using the perfused whole liver or isolated intact hepatocytes are not so well defined as those using homogenates or lysosomal extracts.

The proteolytic activity of parenchymal cells decreased by 40% after treatment with leupeptin or liposomes containing leupeptin. Treatment with pepstatin caused no inhibition of proteolysis because it does not enter the intact cells, whereas liposomes containing pepstatin brought on an inhibition of 25% (Ref. 14).

Similar results using leupeptin not within liposomes have been obtained by Ballard (Ref. 15).

Dean used (Ref. 16) liposomes containing pepstatin and detected a 50% decrease of proteolysis in the perfused whole liver. In addition to hepatocytes, Kupffer cells are also involved in general proteolysis in these experiments. The Kupffer cells contain much more cathepsin D than hepatocytes and therefore pepstatin is more effective. Furthermore it must be born in mind that an increased rate of proteolysis occurs in perfused liver preparations.

Assuming that the lysosomes are involved in intracellular proteolysis, a major difficulty arises in explaining the observed heterogeneity in protein half-lives in terms of autophagic activity. We shall try to assess of the contribution made by the lysosomal proteinases to the selective degradation of cytosol proteins. First, we must note some special properties of short-lived proteins. It is well established (Ref. 17, 18) that most (but not all) short-lived proteins have higher molecular weights as well as more acidic isoelectric points, more superficial hydrophobic areas and in some cases, more wrongly-placed amino acids.

In our opinion a considerable proportion of apolar areas at the surface of the protein molecule is an important factor in the selective catabolism of short-lived proteins. Several steps may be required to generate the superficial hydrophobic areas. In such nonproteolytic primary steps various changes may occur: very large proteins may become dissociated and from monomers, proteins may become dissociated from organelles or from substrates or cosubstrates, and disulfide bonds in proteins may be formed or split. These processes are reversible and therefore they do not determine the rate of degradation in the absence of subsequent reactions, which only occur with the original protein after dissociation. It has been shown that hydrophobic proteins especially tend to aggregate with each other and with membranes. So, different degrees of exposure of superficial hydrophobic areas on substrate protein molecules could cause differing affinities for lysosomal membranes and for lysosomal proteases such as cathepsin L, possibly located on the surface of these membranes. Thus it is not necessary to postulate a special degradation procedure for each of the very many different enzymes with their widely differing half-lives. However, this general mechanism does not exclude the existence of special reactions for the inactivation and degradation of particular proteins.

## REFERENCES

(1) P. Bohley, H.Kirschke, J.Langner, S.Ansorge, H.Hanson, Intrazellulärer Proteinabbau III. Intrazelluläre Verteilung des Zytosolproteinabbaues bei neutralem pH, Acta biol.med.germ. 27, 229 (1971).

(2) D.J.Reijngoud, J.M.Tager, Measurement of intralysosomal pH, Biochim.Biophys.Acta 297, 174 (1973).

(3) P.Bohley, C.Miehe, M.Miehe, S.Ansorge, H.Kirschke, J.Langner, B.Wiederanders, Intrazellulärer Proteinabbau V. Bevorzugter Abbau kurzlebiger Zytosolproteine durch Lysosomenendopeptidasen, Acta biol.med. germ. 28, 323 (1972).

(4) N.Katunuma, E.Kominami, K.Kobayashi, Y.Banno, K.Suzuki, K.Chichibu, Y.Hamaguchi, T.Katsunuma, Studies on New Intracellular Proteases in Various Organs of Rat, Eur.J.Biochem. 52, 37 (1975).

(5) S.Ansorge, H.Kirschke, J.Langner, B.Wiederanders, P.Bohley, H.Hanson, in: Intracellular Protein Catabolism II, ed. by V.Turk, N.Marks, Plenum Press, New York and London, 163 (1977).

(6) J.Langner, S.Ansorge, P.Bohley, H.Welfle, H.Bielka, Presence of an endopeptidase activity in rat liver ribosomes, Acta biol.med.germ. 36, 1729 (1977).

(7) P.Bohley, H.Kirschke, J.Langner, S.Ansorge, Präparative Gewinnung hochgereinigter Lysosomenenzyme aus Rattenlebern, FEBS Lett. 5, 233 (1969).

(8) P.Bohley, H.Kirschke, J.Langner, S.Ansorge, B.Wiederanders, H.Hanson, in: Tissue Proteinases, ed. by A.J.Barrett, J.T.Dingle, North-Holland, Amsterdam, 187 (1971).

(9) K.Otto, Über ein neues Kathepsin, Hoppe-Seyler's Z. Physiol.Chem. 348, 1449 (1967).

(10) A.J.Barrett, Human Cathepsin $B_1$, Bioch.J. 131, 809 (1973).

(11) H.Kirschke, J.Langner, B.Wiederanders, S.Ansorge, P.Bohley, Cathepsin L. A new Proteinase from Rat-Liver Lysosomes, Eur.J.Biochem. 74, 293 (1977).

(12) H. Kirschke, J.Langner, B.Wiederanders, S.Ansorge, P.Bohley, H.Hanson, Cathepsin H. An endoaminopeptidase from rat liver lysosomes, Acta biol.med.germ. 36, 185 (1977).

(13) S.Ansorge, H.Kirschke, K.Friedrich, Conversion of proinsulin into insulin by cathepsins B and L from rat liver lysosomes, Acta biol.med.germ. 36, 1723 (1977).

(14) P.Bohley, S.Riemann, R.Koelsch, J.Lasch, Protein degradation in rat liver cells, Acta biol.med.germ. 36, 1821 (1977).

(15) M.F.Hopgood, M.G.Clark, F.J.Ballard, Inhibition of Protein Degradation in Isolated Rat Hepatocytes, Biochem.J. 164, 399 (1977).

(16) R.T.Dean, Direct evidence of the importance of lysosomes in the degradation of intracellular proteins, Nature 257, 414 (1975).

(17) A.L.Goldberg, J.F.Dice, Intracellular Protein Degradation in Mammalian and Bacterial Cells, Ann.Rev.Biochem. 43, 835 (1974).

(18) A.L.Goldberg, A.C.St.John, Intracellular Protein Degradation in Mammalian and Bacterial Cells, Ann.Rev.Biochem. 45, 747 (1976).

SESSION III

# Turnover of Organelles

# BIOGENESIS AND TURNOVER OF PEROXISOMES

P.B. Lazarow, M. Robbi, and C. de Duve
The Rockefeller University, New York, N.Y. 10021, USA.

Most of the information on the biogenesis and breakdown of peroxisomes comes from investigations on rat liver catalase. This enzyme is made of four identical heme-containing subunits of molecular weight approximately 60,000.

## CHEMICAL PATHWAY OF CATALASE BIOSYNTHESIS

The biosynthesis of both the protein and the heme parts of catalase has been studied on rats receiving by intraportal injection pseudo-pulses of $^{3}$H-leucine, of $^{3}$H-δ-aminolevulinic acid (ALA), or of a mixture of $^{3}$H-leucine with $^{14}$C-ALA (1,2). Biosynthetic products were isolated from the livers by immunoprecipitation with purified anti-catalase, according to a procedure carefully designed to minimize contamination by other proteins. This precaution was very important because catalase is a minor component of liver (less than 0.5% of the total proteins) and has a relatively slow turnover (half-life = 1 1/2 days). Consequently even small amounts of a rapidly turning over impurity would cause serious errors in the estimation of the true radioactivity of catalase and its precursors in immunoprecipitates.

These experiments revealed that the radioactive material precipitated by anticatalase at early times after the injection of labeled leucine differs from authentic catalase in two important respects:
1) It does not survive the standard purification procedure for catalase, and is probably destroyed at the first step by Tsuchihashi's procedure (3) of shaking with a mixture of chloroform and ethanol.
2) Its sedimentation coefficient is of the order of 4S, corresponding roughly to the molecular size of a monomeric subunit of about 60,000 molecular weight.

This catalase precursor has a transient existence, characterized by a turnover time of the order of one hour, and is quantitatively converted into a Tsuchihashi-resistant product, believed to be the completed catalase molecule.

Labeled ALA also appears first in a Tsuchihashi-labile form of catalase, which, however, both appears later and is converted to a resistant form faster, than does the leucine-labeled precursor.

Kinetic analysis of these results led to the conclusion that catalase biosynthesis proceeds by way of at least two distinct labile intermediates, one devoid of heme and of monomeric size, the other containing heme, and presumably also monomeric, although this latter point has not been established. Fig. 1 summarizes this scheme, together with the estimated turnover times and pool sizes of the two intermediates.

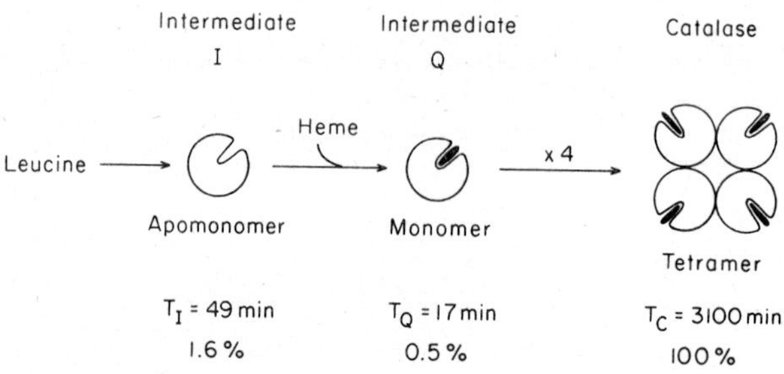


Fig. 1 Biochemical pathway of catalase synthesis (from ref.1). The T's are turnover times and the values underneath them are pool sizes expressed as percentages of the pool size of catalase. The representation of the heme-containing intermediate as a monomer is hypothetical.

More recently, catalase antigen has been made "in vitro" by protein-synthesizing systems derived from either rabbit reticulocytes or wheat germ, and provided with total rat liver polysomal RNA as the source of catalase mRNA (4). The immunoprecipitated product of this synthesis was found to migrate distinctly more slowly upon SDS-polyacrylamide gel electrophoresis than did purified catalase. This difference, illustrated in Fig. 2, corresponds to an apparent excess molecular weight of the order of 4,000. As shown in Fig. 2, immunoprecipitated catalase synthesized "in vivo" also migrates more slowly than does purified catalase, and this difference seems unrelated to the age of the molecule, being essentially the same 8 min and 1 day after synthesis.

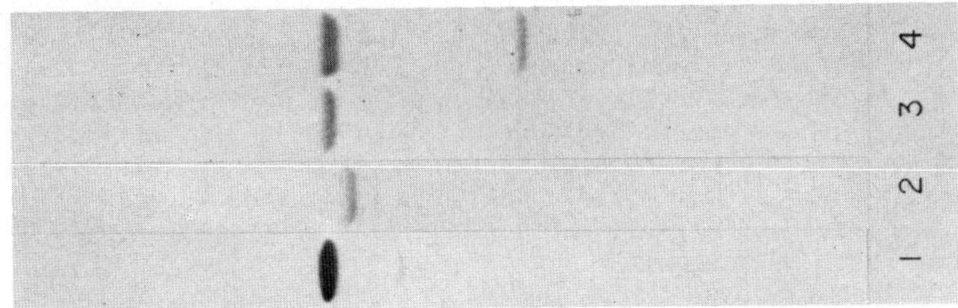


Fig. 2 Fluorogram of SDS gel electrophoresis of catalase species synthesized "in vitro" (Lane 1) and "in vivo" (Lanes 2-4). Lane 1: wheat germ product. 2: purified 1 day old peroxisomal catalase. 3: 8 min old apomonomer precursor immunoprecipitated from high speed supernatant. 4: 1 day old catalase immunoprecipitated from peroxisomes.

The meaning of these results is not yet clear. Synthesis of catalase in the form of a precursor molecule of larger size, followed by proteolytic "processing", seems at first sight not to be involved. On the other hand, until the nature of the effect of the purification procedure on the electrophoretic mobility of SDS-catalase complex is elucidated, no conclusion can be drawn. Of interest in this respect is a recent paper by Walk and Hock (5) describing the "in vitro" translation of mRNA from germinating watermelon cotyledons. Using a specific antibody raised against the malate dehydrogenase containin in the cotyledon glyoxysomes (a particle closely related to animal peroxisomes), these authors isolated a product with an apparent molecular weight, determined by SDS-polyacrylamide gel electrophoresis, of 38,000, as opposed to 33,000 for the purified enzyme.

## INTRACELLULAR PATHWAY OF CATALASE BIOSYNTHESIS

The subcellular location and transport of the biosynthetic intermediates of catalase were investigated by fractionating rat livers 8, 15, 30 and 60 min after the intraportal injection of a mixture of $^{3}H$-leucine and $^{14}C$-ALA (2). Postnuclear supernatants were centrifuged into linear sucrose gradients under conditions that largely resolve the major organelles, partly on the basis of size, and partly on the basis of density.

The results of such an experiment are illustrated in Fig. 3. Two peaks of protein are visible. That at the left, corresponding to the position of the sample layer before centrifugation, consists mainly of the cell sap proteins. The peak in the middle of the gradient is due to the mitochondria (detected by their cytochrome oxidase activity), which are banded at their equilibrium density of 1.18. The position of the peroxisomes at their equilibrium density of 1.23 near the bottom of the gradient is indicated by an arrow in Fig. 3; however, the peroxisomes constitute only 2.5% of the total liver proteins and thus do not form a visible peak in the protein distribution. Under the experimental conditions used, the microsomes (esterase distribution) sediment out of the starting layer, but do not reach their equilibrium position.

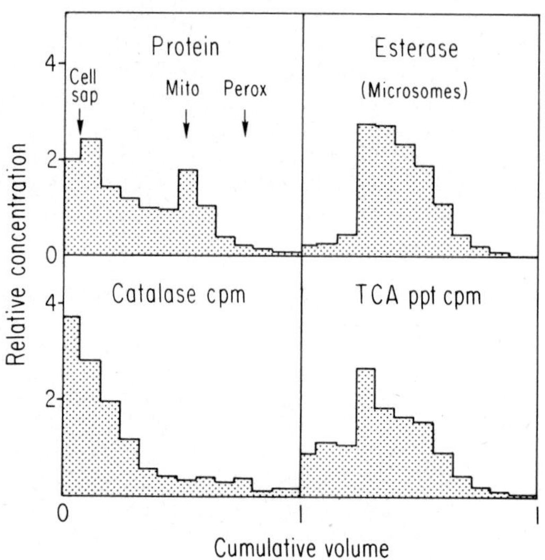


Fig. 3 Subcellular distribution of radioactive catalase and total proteins 8 min after biosynthesis. A postnuclear supernatant of rat liver was centrifuged into a linear sucrose gradient (1.10-1.27 g/cm$^3$) for 1.4x10$^{10}$ rad$^2$/s (consult ref.2 for details). The distributions of total (unlabeled) protein and esterase are given in the top half of the figures; arrows indicate the positions of the cell sap proteins, mitochondria, and peroxisomes. Centrifugation was from left to right. The distributions of radioactivity derived from $^3$H-leucine in catalase immunoprecipitates and in trichloroacetic acid precipitates are shown in the bottom two panels. In this experiment the incorporation of $^3$H-leucine into catalase was approximately 0.3% of that into total proteins.

In the experiment illustrated in Fig. 3, the liver was removed 8 min after injection of the isotopes. As expected, the bulk of the newly synthesized proteins (lower right corner) have a distribution similar to that of esterase, consistent with their sequestration in the microsomes. In contrast, the bulk of the newly synthesized catalase (lower left corner) remains at the top of the gradient where the sample was applied. These results suggest that catalase may be synthesized on free ribosomes and be released into the cell sap. An alternative possibility is that catalase could be synthesized on, and sequestered within, some fragile vesicle which is broken when the liver is homogenized. In any case, it is clear from these results that catalase is <u>not</u> sequestered together with the secretory proteins.

Experiments like that of Fig. 3, but carried out at longer times after labeling, revealed that the newly synthesized catalase was transported to peroxisomes with a half-time of about 14 min, and that this transport was nearly complete at 1 hour. The uptake of the catalase precursors into peroxisomes is more rapid than is the completion of the biosynthetic process; addition of heme and formation of the tetramer occur inside the peroxisomes (Fig. 4).

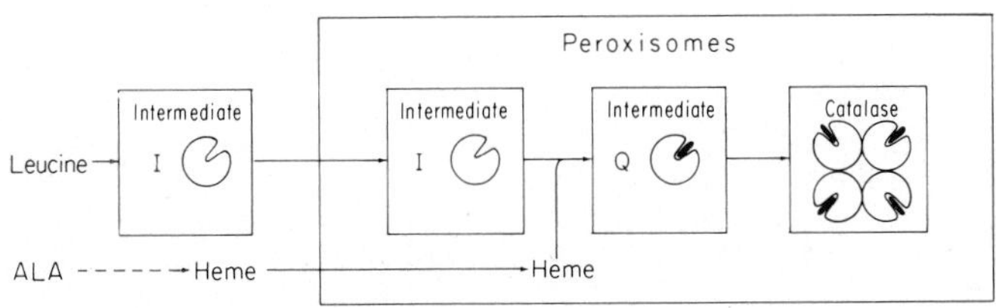


Fig. 4 Intracellular pathway of catalase synthesis (from ref.2).

The mechanism whereby catalase enters the peroxisomes is still mysterious. We have already seen (Fig. 2) that the 8 min old extra-peroxisomal apomonomer precursor and the 1 day old peroxisomal catalase subunit appear to have the same migration upon SDS gel electrophoresis. This does not, however, necessarily exclude the possibility that the precursor contains structural information directing it to the peroxisomes.

REFERENCES

(1) P.B. Lazarow and C. de Duve, J. Cell Biol. 59, 491 (1973).

(2) P.B. Lazarow and C. de Duve, J. Cell Biol. 59, 507 (1973).

(3) M. Tsuchihashi, Biochem. Z. 140, 63 (1923).

(4) M. Robbi and P.B. Lazarow, Proc. Natl. Acad. Sci. U.S.A., in press (1978).

(5) R.A. Walk and B. Hock, Biochem. Biophys. Res. Commun., 81, 636 (1978).

## EFFECT OF OXYGEN ON THE SYNTHESIS OF MITOCHONDRIAL PROTEINS IN SACCHAROMYCES CEREVISIAE

Graeme Woodrow and Gottfried Schatz
Biocenter, University of Basel,
CH-4056 Basel, Switzerland

### ABSTRACT

Oxygen has no significant direct effect on the synthesis of the three mitochondrially-made cytochrome *c* oxidase subunits; however, it causes these subunits to associate with a cytoplasmically-synthesized subunit.

### INTRODUCTION

The formation of a mitochondrial cytochrome system in yeast is dependent on oxygen (1). In order to study the mechanism of this oxygen effect we investigated the influence of oxygen on the pulse-labeling of the mitochondrially-made subunits of cytochrome *c* oxidase in intact yeast cells as well as in isolated mitochondria. In addition, we determined the effect of oxygen on the association of the mitochondrially-made cytochrome *c* oxidase subunits with cytoplasmically-made subunits.

### RESULTS

In order to test for an effect of oxygen on the pulse-labeling of the mitochondrially-made cytochrome *c* oxidase subunits, the haploid strain D273-10B (ATCC 24657) was grown for about 8 generations under strictly anaerobic conditions (2,3) in an all-glass fermenter. When the culture had reached the end of the logarithmic phase, it was chilled and poisoned anaerobically with cycloheximide (100 µg/ml). The cells were washed in the presence of cycloheximide and divided into two aliquots. One aliquot was labeled for 15 min with 35-S-methionine and chased for 20 min with unlabeled methionine in the absence of oxygen. The second aliquot was treated identically except that labeling and chase were done in the presence of oxygen. Labeling and chase were carried out in the presence of 100 µg cycloheximide/ml. Since, in yeast, cycloheximide inhibits cytoplasmic but not mitochondrial protein synthesis, only proteins made in mitochondria became labeled under these conditions (4).

The mitochondria were isolated from each aliquot and the total mitochondrially-synthesized proteins were displayed by high-resolution SDS-polyacrylamide gel electrophoresis followed by radioautography (5). The left part of Fig. 1 shows that oxygen does not significantly affect the labeling of any of these proteins. These proteins include the three large subunits (I-III) of cytochrome c oxidase (4).

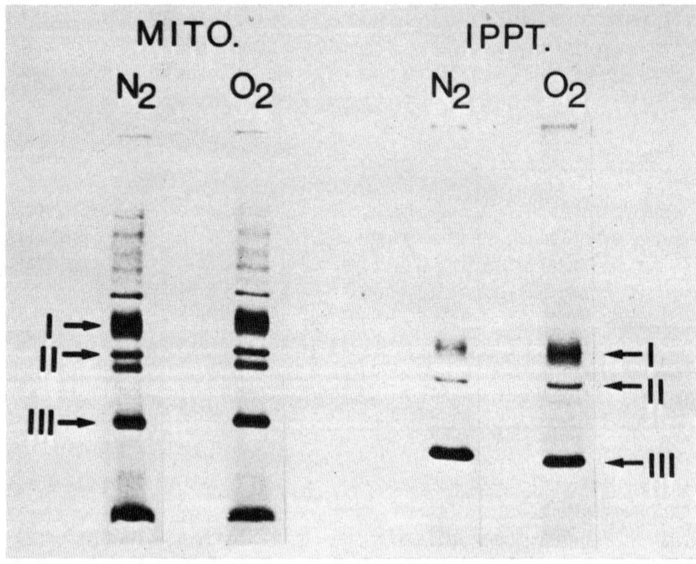


Fig. 1. Oxygen is not necessary for the synthesis of cytochrome c oxidase subunits I-III but stimulates the association of subunits I and II with subunit VI.

However, oxygen causes the pulse-labeled subunits I+ II to associate with at least one of the four cytoplasmically-synthesized subunits. Such an association can be detected with an antiserum specifically directed against the cytoplasmically-made cytochrome c oxidase subunit VI. This antiserum should precipitate subunits I-III only if they are attached to subunit VI. In the experiment depicted in the right half of Fig. 1, the pulse-labeled mitochondria shown in the left half of the Figure were solubilized under conditions (cholate/KCl) which, unlike SDS, do not dissociate active cytochrome c oxidase (6). The extracts were then immunoprecipitated (7) with an antiserum against subunit VI and the immunoprecipitates were analyzed by high-resolution SDS-polyacrylamide gel electrophoresis as described above. It can be seen that whereas subunit III is immunoprecipitated to the same extent from both samples, subunits I and II are precipitated to a far lesser extent in the anaerobically-labeled sample than in the sample labeled aerobically (Fig. 1, right half, and Fig. 2).

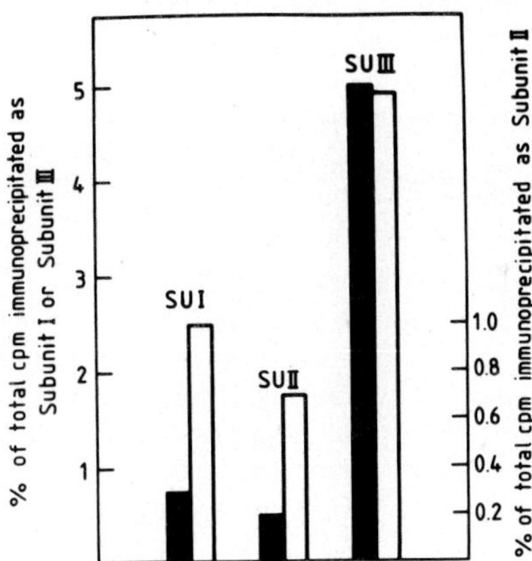


Fig. 2. Quantitative comparison of the relative amounts of subunits I, II and III immunoprecipitated with anti-subunit VI serum from extracts of anaerobically-grown cells labeled either aerobically (open bars) or anaerobically (solid bars). Data were obtained from scans of the autoradiograms presented in Fig. 1.

This "assembly-inducing" effect of oxygen can also be seen if oxygen is only present during the chase (Fig. 3) and if the chase medium is supplemented with sufficient acriflavin to completely block mitochondrial protein synthesis (8). Oxygen does thus not act by affecting either mitochondrial or cytoplasmic protein synthesis.

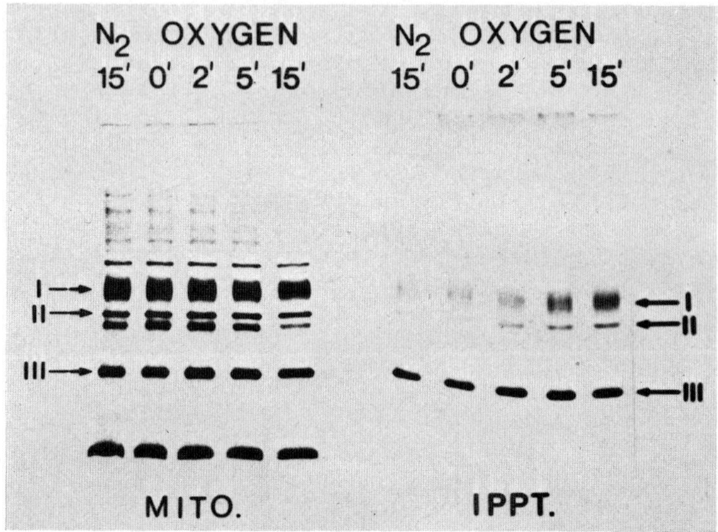


Fig. 3. Effect of oxygen on the association of preformed subunits I-III with subunit VI. Cells were labeled anaerobically with 35-S-methionine at 28°C for 15 min and then chased with unlabeled L-methionine for 15 min. The suspension was then divided into three aliquots. One aliquot (O') was processed immediately as described below. The second aliquot was incubated with vigorous shaking in air at 28°C: the third aliquot was incubated further in the absence of oxygen. Samples of cells were removed at the times shown and promitochondria were isolated. The promitochondria (MITO) as well as the immunoprecipitates (IPPT) obtained from them with antiserum against subunit VI were analyzed by SDS-acrylamide gel electrophoresis and autoradiography.

Closely similar results were obtained with isolated mitochondria. Figure 4 (a-e) shows that respiring mitochondria isolated from aerobically-grown cells can synthesize the same spectrum of polypeptides as mitochondria in cycloheximide-poisoned intact cells (see also ref. 9). It is also shown that a high-speed supernatant greatly stimulates this in vitro system and that, once again, oxygen has no significant effect on pulse-labeling. However, if the in vitro labeled mitochondria are solubilized with cholate/KCl and immunoprecipitated with an antiserum against cytochrome c oxidase subunit VI, coprecipitation of subunits I and II is only observed with the sample that had been labeled in the presence of oxygen (Fig. 4 f-i).

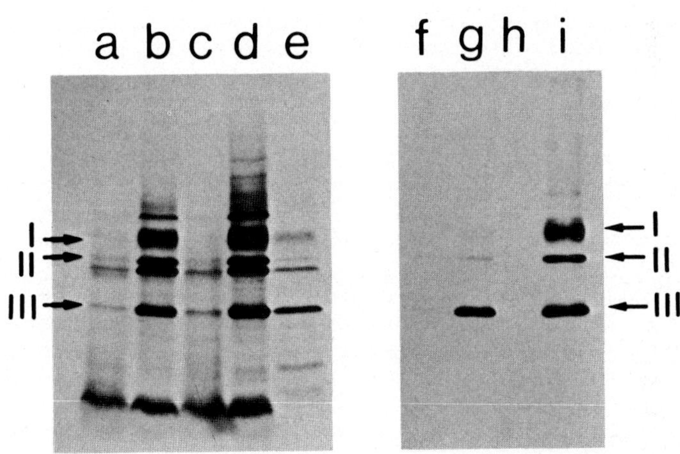


Fig. 4. Effect of oxygen on the synthesis and "assembly" of cytochrome c oxidase subunits in isolated mitochondria obtained from aerobically-grown cells.

a-e: Radioautograms of electrophoretically-resolved mitochondrial translation products labeled with 35-S-methionine.

a: Mitochondria (180 μg) labeled anaerobically.

b: Mitochondria (180 μg) labeled anaerobically in the presence of 470 μg of a high-speed supernatant from aerobically-grown cells.

Fig. 4. c: Mitochondria (180 μg) labeled aerobically.
d: Mitochondria (180 μg) labeled aerobically in the presence of 470 μg of a high-speed supernatant.
e: Mitochondria labeled aerobically in intact cells in the presence of 100 μg/ml cycloheximide.
f-i: Immunoprecipitates obtained from extracts of the mitochondria in a-d (respectively) with antiserum directed against subunit VI.

We conclude that yeast mitochondria can synthesize their major polypeptide products regardless of whether oxygen is present or not. The well-documented oxygen-dependence of cytochrome c oxidase formation does thus not involve an effect of oxygen on the transcription of the mitochondrial genes (10) for subunits I-III or on the translation of the corresponding messenger RNAs. Instead, oxygen appears to trigger association of subunits I-III with their cytoplasmically-made counterparts (IV-VII). How this occurs is still unknown; we are currently exploring the possibility that oxygen causes the formation of heme a which is necessary for the association of the seven cytochrome c oxidase subunits (11).

During long-term anaerobic growth of the yeast cells, the short-term phenomena described here are in part overshadowed by complex regulatory interactions that result in a drastic lowering of all cytochrome c oxidase subunits. Sensitive immunological tests revealed that the concentration of individual cytochrome c oxidase subunits in subcellular fractions from anaerobically-grown cells was 11 to almost 100 fold lower than in the corresponding fractions from aerobically-grown cells (Table I). In contrast, the levels of the three largest subunits of the mitochondrial ATPase varied only by a factor of 4 or less.

TABLE I

Quantitative data were obtained using the immune-replica procedure in conjunction with radio-iodinated protein A, as described in ref. 11.

| Conditions of cell growth | Sample | ng of subunit present in 1 mg of sample | | | | |
|---|---|---|---|---|---|---|
| | | SU II | SU IV | SU V | SU VI | SU VII |
| Anaerobic | Homogenate | -[a] | 26(5)[b] | 11(2.5)[b] | 30(9)[b] | 4(3)[b] |
| | Promitochondria | 20(1)[b] | 15(1)[b] | 140(8)[b] | 140(8)[b] | 140(8)[b] |
| | "Soluble"fraction | -[a] | 25(8)[b] | -[a] | 15(6)[b] | -[a] |
| Aerobic | Homogenate | 100 | 520 | 450 | 350 | 120 |
| | Mitochondria | 1700 | 1350 | 1850 | 1700 | 1800 |
| | "Soluble"fraction | -[a] | 320 | -[a] | 240 | -[a] |

[a] Not detected

[b] Value in parenthesis is the amount of each subunit present in anaerobically-grown cells expressed as a percentage of that present in the corresponding fraction of cells grown aerobically.

These long-term oxygen effects may be related to the observation that the synthesis of polypeptides in mitochondria is stimulated by the availability of cytoplasmically-made "partner proteins" (6). If this stimulation requires proper association between the partner polypeptides, then it would be oxygen dependent since our present data show that the association of subunits I and II with subunit VI requires oxygen.

## REFERENCES

(1) P.P. Slonimski, La formation des enzymes respiratoires chez la levure, Masson, Paris 1953.
(2) G. Schatz and L. Kováč, Methods Enzymol. 31, 627 (1974).
(3) G. Woodrow, unpublished.
(4) G. Schatz and T.L. Mason, Annu. Rev. Biochem. 43, 51 (1974).
(5) M.G. Douglas and R.A. Butow, Proc. Natl. Acad. Sci. USA 73, 1083 (1976).
(6) T.L. Mason, R.O. Poyton, D.C. Wharton and G. Schatz, J. Biol. Chem. 248, 1346 (1973).
(7) E. Ebner, T.L. Mason and G. Schatz, J. Biol. Chem. 248, 5369 (1973).
(8) G.S.P. Groot, W. Rouslin and G. Schatz, J. Biol. Chem. 247, 1735 (1972).
(9) G.S.P. Groot and R.O. Poyton, Nature 255, 238 (1975).
(10) F. Cabral, M. Solioz, Y. Rudin, G. Schatz, L. Clavilier and P.P. Slonimski, J. Biol. Chem. 253, 297 (1978).
(11) J. Saltzgaber-Müller and G. Schatz, J. Biol. Chem. 253, 305 (1978).

# DEGRADATIVE PROCESSES IN YEAST MITOCHONDRIA AND THEIR RELATION TO MITOCHONDRIAL BIOGENESIS

Valentin N.Luzikov
A.N.Belozersky Laboratory of Molecular Biology and Bioorganic Chemistry,
Lomonosov State University,
Moscow 117234, USSR

## ABSTRACT

Consideration is given to some phenomena which may be related to regulation of mitochondrial biogenesis in yeast. One of them is the dependence of mitochondrial stability in yeast cells on the functioning of mitochondrial energy-generating and energy-transducing systems. Another issue is the very rapid proteolysis of mitochondrial translation products both *in vivo* and *in vitro*. It is suggested that some non-functioning components of the respiratory system possibly appear at the early stages of mitochondrial formation in yeast, and these components are eliminated in the course of mitochondrial differentiation.

## INTRODUCTION

Degradation of mitochondria or their individual components is an integral part of mitochondrial biogenesis. Degradative processes should not only be considered as a means of renewal of mitochondria and regulation of the mitochondrial mass. We believe these processes to be a creative force controlling mitochondrial assembly and determining their "final state", i.e. composition, stoichiometry between components, ultrastructure etc.

This communication will cover two aspects that may have a direct bearing on the regulation of mitochondrial assembly in yeast. The first one is the significance of oxidative phosphorylation for mitochondrial integrity in yeast cells. The second item is the efficient proteolytic machinery inherent in yeast mitochondria. The last section presents some preliminary data in evidence for proteolytic control of the respiratory system assembly which implies elimination of non-functioning components.

## RESULTS AND DISCUSSION

### A Functional Basis for Mitochondrial Integrity

The dependence of mitochondrial *in vivo* integrity on their functional state can be best demonstrated with facultatively anaerobic organisms, such as *Saccharomyces cerevisiae*. This yeast is known to complete the development of its respiratory machinery by the early stationary phase of aerobic growth in a glucose-containing medium. Deaeration of such cells entails a decrease in the specific activities of the respiratory chain in isolated mitochondria. The most rapidly decreasing are NADH oxidase and succinate oxidase, while some partial activities are only slightly changed /1/. As follows from Fig. 1, similar alterations in the respiratory chain are induced in aerobic conditions by a respiratory inhibitor, cyanide, and uncouplers, 2,4-dinitrophenol /DNP/, and carbonylcyanide-m-chlorophenyl hydrazone /CCCP/. A close correspondence between the effects of the above agents with that of anaerobiosis is likely to mean that the fundamental cause of the degradation of the respiratory chain in all cases lies in the interruption of oxidative phosphorylation /2/. The decrease in the specific respiratory activities of the mitochondrial fraction is accompanied by degradative alterations in organelle morphology, as revealed electron microscopically. Moreover, the morphological effects of the uncouplers and anaerobiosis appear to be rather similar /3/.

The set of data available at the present moment gives grounds for discerning two basic mechanisms regulating the degradation of yeast mitochondria. One of them supposes that the degradation is triggered by the induction of synthesis of lytic enzymes. Grossman et al. /4/ and Dharmalingam and Jayaraman /5/ suggested that the degradation of mitochondria during glucose repression in yeast is associated with the induction of phospholipase D. Similarly, Stone and Wilkie /6/ believe a cytoplasmically synthesized peptidase to be responsible for the degradation of mitochondrial cytochrome oxidase upon addition of chloramphenicol to growing *S.cerevisiae*. Indeed, in both cases degradation was suppressed by cycloheximide, an inhibitor of cytoplasmic translation. This mechanism, however, can hardly be applied to deaeration of yeast in the early stationary phase. As can be seen from Fig. 1, the inhibitors of protein synthesis, chloramphenicol /CAP/ and cycloheximide /CHI/ do not prevent degradation of the respiratory system induced by anaerobiosis.

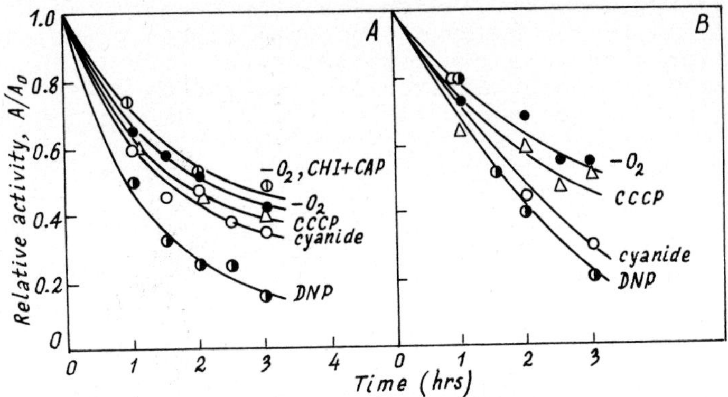


Fig. 1. Decrease in NADH oxidase and succinate oxidase activities of mitochondria upon incubation of aerobically grown yeast /early stationary phase/ in anaerobic conditions or in the presence of cyanide /3 mM/, DNP /0.1 mM/, or CCCP /1 µM/. Deaeration of yeast in the presence of CHI /25 µg/ml/ and CAP /4 mg/ml/. At times indicated samples of cell suspension were withdrawn, mitochondria were isolated and the activities indicated were assayed. $A/A_0$ is the ratio of the current to initial activity. A, NADH oxidase; B, succinate oxidase.

Another regulatory mechanism can be proposed on the basis of data concerning the dependence of the stability of the oxidative phosphorylation system on its functional state. A series of model experiments in our laboratory /7/ has demonstrated that this system can withstand thermal inactivation and the action of proteinases and phospholipase A when it performs electron transfer coupled with ADP phosphorylation. These data allow a suggestion to be made that the described degradation of yeast mitochondrial respiratory system was triggered by the destabilization of the latter owing to interrupted oxidative phosphorylation. Questions to be further clarified are the degree of degradation and the particular destructive processes induced during anaerobiosis or in the presence of respiratory inhibitors or uncouplers /proteolysis, lipolysis, thermal disintegration of the membrane at physiological temperatures etc./.

Thus, the integrity of yeast mitochondria is determined by the fact whether the cell performs oxidative phosphorylation. Speaking of a possible relation of the above phenomena to regulation of mitochondrial formation, one should mention that the most sensitive to interruption of oxidative phosphorylation are mitochondria of the yeast cells in the early exponential growth phase. These mitochondria possess low respiratory activities which completely disappear after a 3-h deaeration of the yeast suspension /3/, as depicted in Fig. 2. The mentioned data

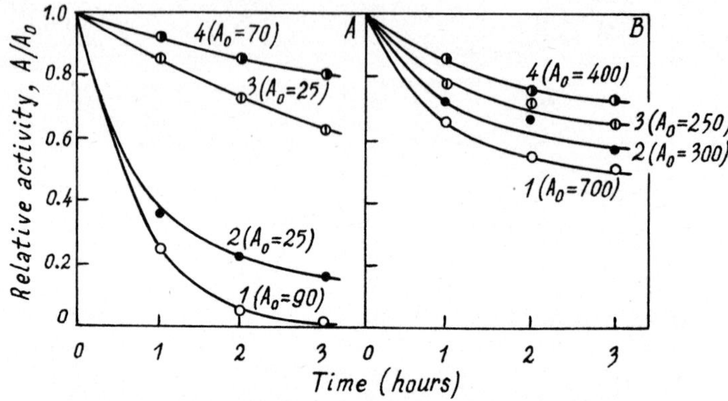


Fig. 2. Degradation of mitochondrial respiratory system upon deaeration of yeast cells at the early exponential /A/ and early stationary /B/ growth phases. Yeast cell suspensions were thickened after 7.5 h or 13 h of aerobic growth to 5 per cent /wet weight/and incubated in anaerobic conditions. Other details as indicated in Fig. 1. Curve 1, NADH oxidase; 2, succinate oxidase; 3, cytochrome c oxidase; 4, succinate:ferricyanide reductase. Activities in brackets are expressed as nmoles of substrate oxidized per min per mg of mitochondrial protein.

suggest that conditions existing in young yeast cells are especially favourable for disorganization of non-functioning mitochondria or parts thereof, if such may appear during the initial stages of respiratory system formation.

## Endogenous Proteolysis of Mitochondrial Translation Products

The second point which may be of crucial importance for regulation of formation of mitochondria is that the latter possess endogenous proteinases digesting mitochondrial translation products. According to the concepts of Tzagoloff et al. /8/, Ebner et al. /9/ and Werner et al. /10/, the products of mitochondrial protein synthesis play an organizing role in the assembly of enzymatic complexes of the mitochondrial inner membrane. This fact makes obvious the importance of information on the half-lives of the above-mentioned proteins. Such information has until recently been lacking owing to a number of experimental problems.

A special method for the determination of the degradation rates of mitochondrial translation products in yeast was developed in our laboratory on the basis of double-labelling pulse-chase techniques /11/. The experimental conditions ensured that recycling of the label was suppressed and its post-incorporation into mitochondrial protein could be accounted for. According to the data obtained /11/, the half-life of [$^3$H] leucine incorporated into mitochondrial translation products in the presence of cycloheximide is about 60 min in the early stationary phase of yeast growth. As Fig. 3 demonstrates, the loss of the label from protein is suppressed by phenylmethyl sulfonylfluoride /PMSF/ and leupeptin /LP/ but insensitive to pepstatin /PS/. Since leupeptin is not inhibitory towards yeast cytoplasmic proteinases A, B and C, according to Lenney /12/, it can be suggested that the proteolysis of mitochondrial translation products in yeast cells is at least partially associated with the action of an endogenous mitochondrial proteinase.

This suggestion was further confirmed by studies on the proteolysis of mitochondrial translation products in isolated yeast mitochondria. Recently published data /12/ indicate that proteolysis of mitochondrial translation products proceeds in vitro at a high rate comparable with that observed in vivo. Contamination with cytoplasmic proteinases can be almost certainly ruled out since the mitochondrial fraction displayed no activity in standard assays for cytoplasmic proteinases A and B, which appear to be the main enzymes responsible for degradation of the total cell protein. Moreover, the breakdown of mitochondrial translation products was insensitive to antibodies against these proteinases.

Thus yeast mitochondria in all probability possess an endogenous proteinase /or proteinases/ which can efficiently digest mitochondrial translation products at neutral pH. Table 1 summarizes data on the inhibitor sensitivity of the mitochondrial proteolytic activity. The latter can be ascribed to a serine protease, judging from its inhibition by phenylmethylsulfonylfluoride. Yet, a sulfhydryl reagent, p-chloromercuribenzosulphonate /p-CMBS/, is also inhibitory. A characteristic feature of the mitochondrial enzyme is its sensitivity to leupeptin which has no effect on any of the cytoplasmic proteinases A, B and C /13/.

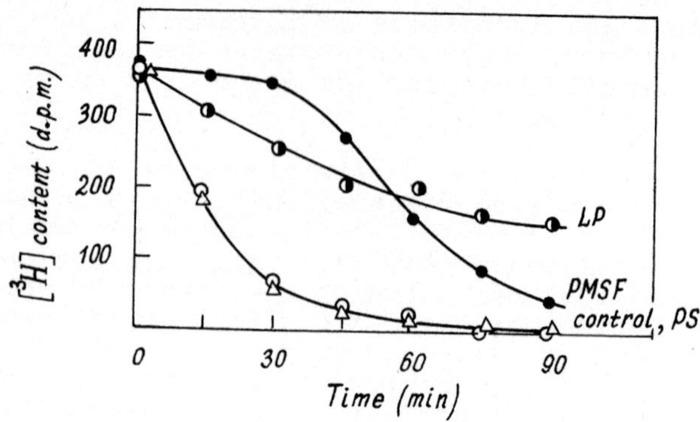


Fig. 3. Effect of proteinase inhibitors on the decrease in the radioactivity of the pulse-labelled mitochondrial translation products upon incubation of aerobically grown yeast at the early exponential phase. PMSF /50 µmoles per litre/ was introduced into the growth medium 10 min before the end of the pulse, at zero and 10 min of the chase. LP was added at a concentration of 2 mg per litre at the beginning of the chase.

Correlating the effects of various inhibitors on cytoplasmic proteinases A and B with those on the proteolysis of mitochondrial translation products in isolated mitochondria and in intact cells, we can conclude that a leupeptin-sensitive intramitochondrial proteinase is at least partly responsible for the <u>in vivo</u> proteolysis of the mentioned proteins.

Comparison of the data presented in /11/ demonstrates that the rate of proteolysis in the mitochondria of the exponentially growing yeast cells possessing an imperfect respiratory system is significantly higher than in the fully competent mitochondria of the cells in the early stationary phase. These data mean that during the initial stages of the development of the respiratory system, yeast mitochondria have a remarkable capacity to degrade polypeptides synthesized on mitoribosomes. Fig. 3 indicates that the half-life of such polypeptides may be as short as 20 min.

All the above-said leads to the following conclusions. Firstly, yeast mitochondria possess an efficient machinery for elimination of the products of mitochondrial protein synthesis. Since the polypeptides under conside-

TABLE 1, Inhibition of the activities of proteinases A and B and intramitochondrial proteolysis

| Inhibitor | Concentration | Inhibition, per cent | | |
|---|---|---|---|---|
| | | Proteinase A | Proteinase B | Intramitochondrial proteolysis |
| PMSF | 0.2 mM | 0 | 100 | 76 |
| | 1.0 mM | 0 | 100 | 85 |
| p-CMBS | 1.0 mM | 0 | 100 | 75 |
| | 2.0 mM | 0 | 100 | 89 |
| Pepstatin | 1.0 µg/ml | 100 | 0 | 0 |
| Leupeptin | 2.0 µg/ml | 5 | 0 | 55 |
| Antipain | 2.0 µg/ml | 0 | 100 | 95 |
| Chymostatin | 2.0 µg/ml | 0 | 100 | 83 |
| Chloramphenicol | 4.0 mg/ml | 0 | 0 | 95 |

ration were produced in the absence of cytoplasmic translation, it can be thought that the observed exceptionally short half-lives are characteristic of the proteins that are not properly integrated into the mitochondrial membrane owing to the lack of cytoplasmic "partners". Secondly, even the competent respiratory chain rapidly degrades if for some reasons its functioning is interrupted.

The question arises whether such situations can occur during development of the yeast respiratory system under normal conditions when oxidative phosphorylation takes place and both protein-synthesizing systems are coordinated. In other words, are single polypeptides or whole enzymatic complexes apt to be improperly integrated into the membrane and preferentially degraded therefore? Some preliminary data of our laboratory which are presented below testify that the answer may be positive.

## The Development of the Cell Respiratory System and the Effect of Proteinase Inhibitors

The development of the respiratory system of intact S.cerevisiae cells was studied during aerobic growth in a galactose medium /14/. As shown in Fig. 4, the cell respiration reached its maximal value in the early exponential phase of culture growth and then remained constant. The cell respiration apparently reflected the respiratory activity of the mitochondrial population, since it was stimulated by an uncoupler /carbonylcyanide-m-chlorophenyl hydrazone/ and completely inhibited by cyanide and antimycin. At the same time the cell content of cytochromes $\underline{aa}_3$ $\underline{b}$, $\underline{c}$ and $\underline{c}_1$ reached a maximum simultaneously with the respiratory activity, but then gradually decreased through the exponential phase.

These results imply synthesis of a certain amount of cytochromes which are not necessary for the maintenance of a constant respiration level. It should be noted that these "excess" cytochromes are completely reduced when the cells reach anaerobiosis by respiration on glucose or ethanol, and the stoichiometry between individual components is very close to that observed in the cells from the early stationary phase. No unusual components absorbing in the cytochrome spectral region could be detected.

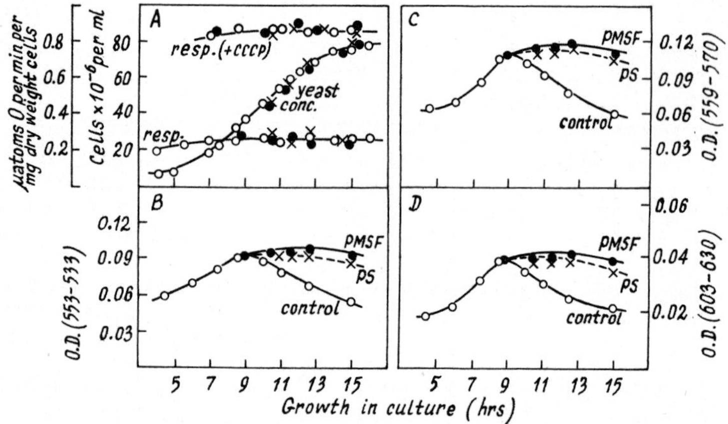


Fig. 4. Effect of proteinase inhibitors on aerobic growth of Saccharomyces cerevisiae in galactose-containing medium, cell respiration and cytochrome content.
-o-o-, no additions; -●-●-, in the presence of phenylmethylsulfonylfluoride /PMSF, cumulative 0.2 mM/; -x-x-, in the presence of pepstatin /PS, 1 mg/l cell suspension/.
A, growth in culture:curve 1, concentration of cell suspension; curve 2, nonstimulated respiration on 2% galactose; curve 3, maximal CCCP-stimulated respiration on 3% /v/v/ ethanol. B-D, cytochrome content of the intact cells. The data are presented in form of absorbance of each cytochrome in the difference /reduced minus oxidized/ low-temperature spectra of cell suspensions containing 30 mg dry weight per ml /1.2 x $10^9$ cells per ml/.

In principle, it could be thought that a certain portion of a particular cytochrome may bear no relation to the respiratory chain, but perform some other yet unknown functions. However, considering that this "excess" cytochrome moiety comprises all types of cytochrome components which are /i/ spectrally indistinguishable from "normal", /ii/ present in usual stoichiometry and /iii/ readily reducible by physiological substrates, such an assumption seems hardly probable. A more plausible explanation is that these cytochromes are somehow associated with the respiratory chains but, for unknown reasons, do not participate in cell respiration.

The decrease in the cell cytochrome content could be due either to their elimination or "dilution" during cell divisions. The data concerning the effect of proteolytic inhibitors on this process seem to support the first alternative. As should be apparent from Fig. 4, phenylmethylsulfonylfluoride, an inhibitor of proteinase B /15/ and the intramitochondrial proteinase /see above/, as well as pepstatin which selectively inhibits proteinase A /15/, prevent or attenuate the decrease in the cell cytochrome content during culture growth. Meanwhile no effect can be observed on cell respiration either in the absence or in the presence of an uncoupler.

Thus the above data suggest that the development of the cell respiratory apparatus during yeast aerobic growth in a galactose medium implies synthesis of a certain "excess" amount of cytochromes which are then gradually eliminated without any decrease in the overall respiratory activity. A possible explanation of this fact is that "excess" cytochromes represent the non-functioning membrane components which are eliminated owing to their instability. This viewpoint, however, needs further substantiation, and studies along this line are under way in our laboratory.

## REFERENCES

/1/ V.N. Luzikov, A.S. Zubatov, E.I. Rainina, L.E. Bakeyeva, Biochim. Biophys. Acta 245, 321-334, 1971.

/2/ V.N. Luzikov, A.S. Zubatov, E.I. Rainina, FEBS Letters 11, 233-236, 1970.

/3/ V.N. Luzikov, A.S. Zubatov, E.I. Rainina, J. Bioenerg. 5, 129-149, 1973.

/4/ S. Grossman, J. Cobley, P.K. Hogue, E.B. Kearney, T.P. Singer, Arch. Biochem. Biophys., 158, 744-753, 1973.

/5/ K. Dharmalingam, J. Jayaraman, Biochem. Biophys. Res. Commun., 45, 1115-1118, 1971.

/6/ A.B. Stone, D. Wilkie, J. Gen. Microbiol., 91, 150-156, 1975.
/7/ V.N. Luzikov, Sub-Cell. Biochem., 2, 1-31, 1973.
/8/ A. Tzagoloff, J. Biol. Chem., 246, 3050-3056, 1971.
/9/ E. Ebner, T.L. Mason, G. Schatz, J. Biol. Chem., 248, 5369-5378, 1973.
/10/ S. Werner, A.J. Schwab, W. Neupert, Eur. J. Biochem. 49, 607-617, 1974.
/11/ G.Ya. Bakalkin, S.L. Kalnov, A.V. Galkin, A.S. Zubatov, V.N. Luzikov, Biochem. J., 170, 569-576, 1978.
/12/ S.L. Kalnov, N.V. Serebryakova, A.S. Zubatov, V.N. Luzikov, Biokhimiya /Russ./ 43, 662-667, 1978.
/13/ J.F. Lenney, J. Bacteriol., 122, 1265-1273, 1975.
/14/ V.N. Luzikov, T.A. Makhlis, A.V. Galkin, FEBS Letters 69, 108-110, 1976.
/15/ T. Saheki, H. Holzer, Eur.J. Biochem. 42, 621-626, 1974.

# MITOCHONDRIAL PROTEIN DEGRADATION

S. Grisolía, E. Knecht, J. Cervera and J. Hernández
Inst. Invest. Citológicas, Amadeo de Saboya, 4,
Valencia - 10, Spain

The importance of and need to clarify the mechanism(s) which regulate protein turnover, particularly intracellular degradation of proteins, have been pointed out repeatedly (1). Although there is evidence for alternate systems of intracellular protein breakdown, lysosomes have been frequently proposed as the main proteolytic agents (2). However, direct evidence for lysosomal participation in this process is lacking.

We have concentrated our work in mitochondrial proteins. There are likely several mechanisms of protein degradation and therefore by studying these organelles we should restrict the problem. Most proteins are made in the cytosol and migrate to the mitochondria; either the bulk of these proteins are degraded within the organelles or without in which case they must come out. Alternatively proteases may enter mitochondria or there may be initiation of degradation in mitochondria to be followed by extensive degradation outside.

Mitochondrial enzymes have different half-lives (Table 1).

TABLE 1 Half-lives for some Mitochondrial Proteins from Rat Liver

| Protein | Half-life | Reference |
|---|---|---|
| δ - Aminolevulinate synthetase | 0.3 h | 2 |
| PEP carboxykinase | 5.0 h | 2 |
| Ornithine aminotransferase | 19 h | 2 |
| Alanine aminotransferase | 20 h | 2 |
| Glutamate dehydrogenase | 1 d | 3 |
| Adenosine triphosphatase | 2.5 d | - |
| Malate dehydrogenase | 2.6 d | 3 |
| α - Glycerophosphate dehydrogenase | 4 d | 2 |
| Cytochrome $b_5$ | 5.1 d | 2 |
| Carbamyl phosphate synthetase | 7.7 d | 3 |

Therefore, bulk or unregulated autophagy seems unlikely. Although non-lysosomal proteases specific for pyridoxal-requiring enzymes have been reported in mitochondrial preparations (4), exclusion of lysosomal participation is difficult, indeed mitochondria are usually contaminated with lysosomes, and evidence has been presented pointing to a main role for lysosomes in mitochondrial proteolysis (5,6).

## EXPERIMENTS WITH MITOCHONDRIAL PREPARATIONS

The participation of lysosomal enzymes on the bulk of protein degradation by mitochondrial preparations are typified in Fig. 1-2.

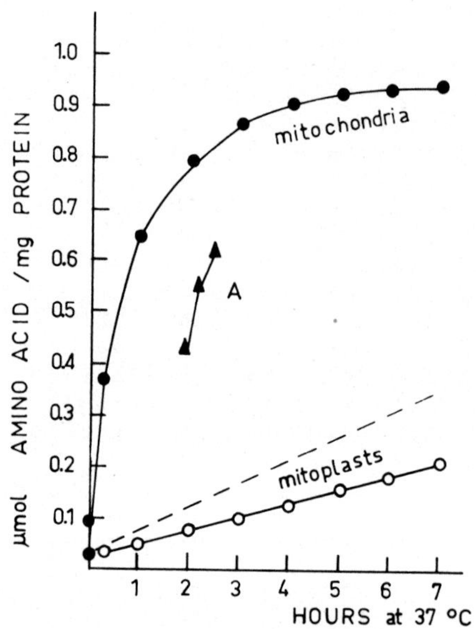


Fig. 1

As shown, there is extensive proteolysis at pH 7.4 with mitochondria (50 mg protein/ml). However, with mitoplast under the same conditions the rate is reduced. Moreover, as shown by the line marked A on addition of mitoplasts to mitochondria which has been aged by two hours of incubation there was an increase in proteolysis. That illustrates that mitoplasts can be substrate for the lysosomal contamination (or that there is cooperation with mitoplast proteolysis; see below). The preparation was examined for lysosomal contaminant. The dotted line shows the calculated proteolysis based on the β-NAGase contamination (∿8%) of that found in mitochondria. This is less than expected. Moreover, electron microscopy revealed negligible contamination with lysosomes. Of course, the presence of larger amount of soluble proteases or attached to membranes, cannot be excluded.

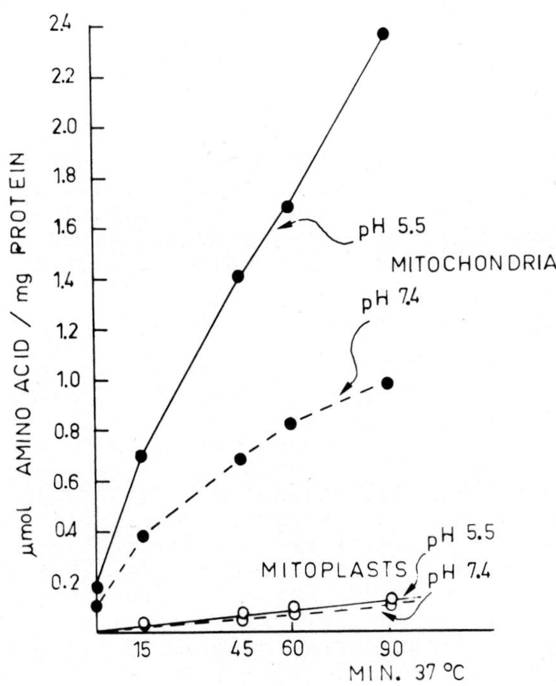


Fig. 2

As depicted in Fig. 2, proteolysis with mitoplasts remains ca. equal at pH's 5.5 and 7.4. With the mitochondrial preparation proteolysis increases ca. 2-fold. However, as illustrated in Table 2, cytosol protein is 140 times more susceptible to proteolysis by lysosomal enzymes at acid pH, (7).

Perhaps the very large increase in proteolysis at low pH with cytosol and other proteins has somewhat misled attention to the high activity at neutral or close to intracellular pH of lysosomes, which may be more physiological. Certainly there is sufficient proteolysis by mitoplast at neutral pH to account for physiological turnover.

TABLE 2 Relative degradation of protein by lysosomal preparations at acid & neutral pH

| | Acid pH | Neutral pH |
|---|---|---|
| Cytosol | 140 | 1 |
| Mitochondria | 2 | 1 |
| Mitoplast | 1 | 1 |

## RADIOAUTOGRAPHY EXPERIMENTS

If lysosomes participate in the degradation of much of the mitochondrial proteins, they could contain these proteins or their hydrolytic products which could be detected.
MS cells, an established monkey kidney epithelium cell line, were cultured (8). Cell growth was measured with a Coulter counter. Exponentially growing cells were incubated for 10 min without and with inhibitors. ($^3$H)-leucine (5 μCi/ml) was added. Cells were incubated for 1 h at 37°C. The well washed cells were precipitated with 10% trichloroacetic acid. The protein was dissolved in 0.2 N NaOH, and the radioactivity determined by liquid scintillation. At the concentrations used, cycloheximide (200 μg/ml) produced only small changes in the fine structure of cells, (Golgi areas were somewhat reduced; almost complete absence of autophagic vacuoles).
The effects of cycloheximide on cell growth were reversible after a lag. The incorporation of ($^3$H)-leucine was inhibited 96.8% with cycloheximide. With cycloheximide plus chloramphenicol, there was a 33.3% inhibition over the cycloheximide-insensitive ($^3$H)-leucine incorporation.
For radioautography, cells were grown for 10 min in the presence of cycloheximide, then ($^3$H)-leucine (57 Ci/mM) was added to give 300 μCi/ml, and after 1 h growth washed exhaustively. Then they were immediately fixed or further incubated (chase) for 3, 6, 12, 24 and 48 h. Electron microscopic radioautography was carried out (8). 50 electron micrographs for each period, obtained at random, were used (final magnification: x 12000). The distribution of silver grains was determined by standard methods.
Radioautographs revealed the majority of silver grains overlaying mitochondria (Fig. 3). The specific activities (# of grains/area) in the compartments revealed that only mitochondria and lysosomes had label significantly higher than the mean value for the cell. Although the % of grains overlaying lysosomes was small, they represent only ∿ 1% of the cell volume. Therefore, their specific activity showed them to be heavily labeled. Since mitochondria represent the major site of protein synthesis in cultured cells incubated with cycloheximide, the specific activity found in lysosomes could signify a selective mechanism for certain mitochondrial proteins.

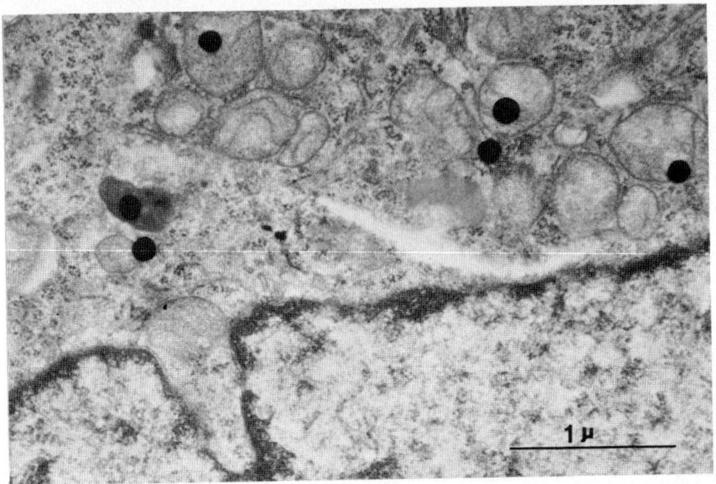


Fig. 3

To clarify these results, decay of radioactivity in mitochondria were plotted semilogarithmically (Fig. 4). The best fitted regression lines indicate two major groups of mitochondrial proteins, i.e. with half-lives ($t_{1/2}$) of 1.6 h and 6.3 d, respectively. Such gross classification has been shown to be valid for mammalian cells.

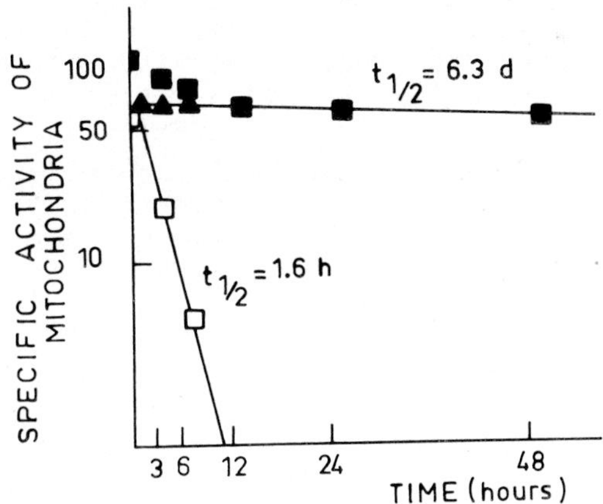


Fig. 4

As shown, (Fig. 5), the decline in radioactivity of mitochondria is followed by a rise of radioactivity of lysosomes (during the first 12 h). Thus, lysosomes seem of importance, at least for the degradation of proteins with short $t_{1/2}$. These results are consistent with others indicating that proteins with short $t_{1/2}$ are more susceptible to lysosomal proteases (2).

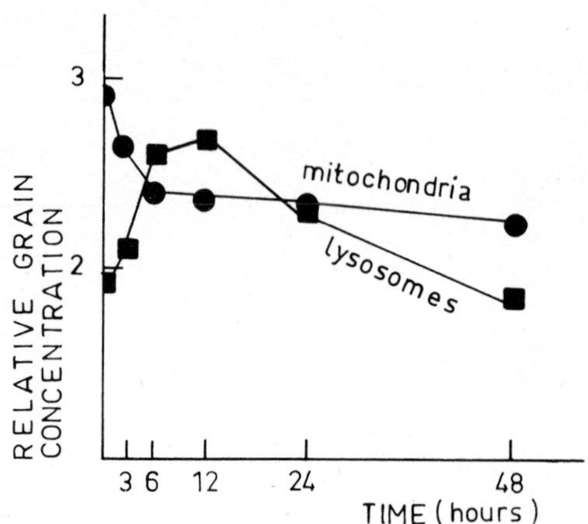


Fig. 5

## IMMUNOCYTOCHEMICAL LOCATION OF CARBAMYL PHOSPHATE SYNTHETASE (CPS) FROM RAT LIVER

Immunocytochemical location of mitochondrial enzymes could, because of its high specificity and sensitivity, be a powerful tool to investigate the site and mechanism of degradation of these enzymes. We started with CPS, a key enzyme in urea synthesis, because it makes ∿ 20% of the protein of mitochondrial matrix of ureotelic liver (9).

CPS was prepared from rat liver mitochondria (6). Antiserum to CPS from rat liver was prepared (10). Activity and purity of antiserum, non-purified and purified IgGs were tested by inhibition of CPS, by precipitin reactions and/or by double immunodiffusion (11). Purified antibodies were conjugated with ferritin using m-xylylene diisocyanate. Tissues or fractions thereof were fixed in 2% paraformaldehyde and 0.2% picric acid buffered with 0.1 M phosphate at 4°C for 30 min and washed overnight in phosphate buffered-saline. The treatment of samples varied depending on which technique was used. For the postembedding procedure, fixed samples of liver were dehydrated in acetone embedded in Epon 812 and polymerized at 40°C. Ultrathin sections were treated with Na-methoxide solution (12). After washing the grids were treated with the ferritin-anti CPS at 22°C. For the preembedding procedure, samples of fixed homogenate and mitochondrial fractions were incubated at 22°C with ferritin-anti-CPS. The samples were then thoroughly rinsed, post-fixed in 2% $OsO_4$ in veronal acetate buffer, dehydrated in acetones and embedded in Epon by standard methods. Controls of normal rabbit IgG labeled with ferritin instead of the ferritin-anti-CPS and with pure ferritin were carried out. Ferritin particles were counted on electron micrographs. CPS was found in the mitochondrial matrix with both techniques (Fig. 6-7).

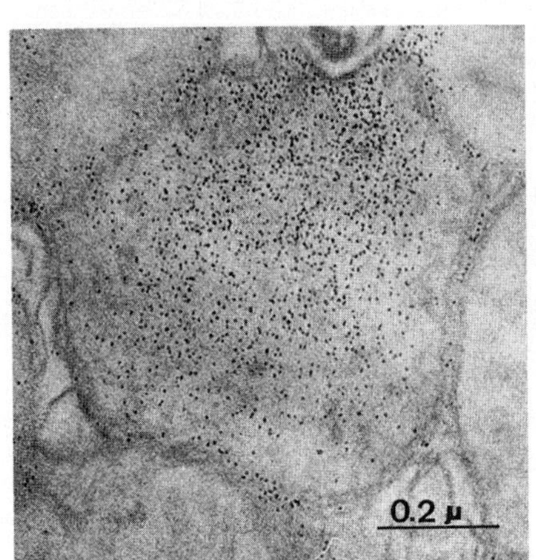


Fig. 6

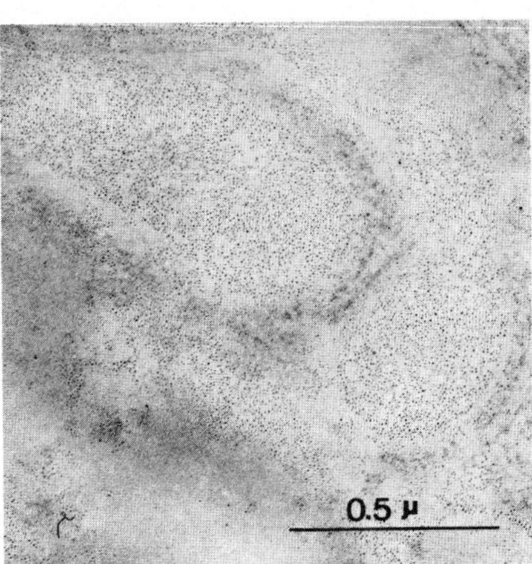


Fig. 7

With subcellular fractions many mitochondria were not labeled, however, with tissue sections all were labeled, possibly due to different accessibility with preembedding versus postembedding techniques. When using the latter technique there was a nonspecific background. With the preembedding technique some nonspecific binding of ferritin conjugates to membranes was also observed. Controls showed no ferritin particles in the mitochondrial matrix. It appears, therefore, that the homogeneously distributed staining by the ferritin-anti-CPS conjugate of the mitochondrial matrix is specific for CPS. Activity measurements (13) indicated that the major portion of CPS is associated with the mitochondrial matrix. Our results show this directly and that the CPS molecules are randomly dispersed in the mitochondrial matrix. The ferritin particles per unit mitochondrial area were 6221.1 ± 621.5 molecules/$\mu^2$. The mean section thickness was similar and the same ferritin conjugates were used through out. These precautions should minimize errors. The mean section thickness can be known only approximately; based on interference colour index scale a thickness of ∿ 800 Å was used. Accordingly, the number of ferritin molecules/$\mu^3$ of mitochondria is ∿ 75,000. Using a mitochondrial volume of 0.65 $\mu^3$ there would be ∿ 48,500 CPS molecules in rat liver mitochondria. Assuming 1 ml of mitochondria equals 1 ml of water, rat liver mitochondria contains 0.12 mM CPS. These values are minimal since obviously there are solids therein and since the cells antigens must be stabilized by fixation to prevent extraction prior to the application of the anti-CPS-ferritin. On the other hand, since the # of ferritin particles is ∿ equal in all cells, it appears that any variation occurs to the same extent. This is highly interesting for and if when the same technique is applied to other mitochondrial enzymes, it might be possible to demonstrate whether or not the immunocytochemical method is superior to other methods thus far available to calculate enzyme concentration. The amount of CPS calculated above is very close to that obtained by others (9). However, it has been recently reported that the CPS of rat liver mitochondria is 1-1.5 mM (14); since, all these values are based on activity measurements (based on extrapolation from the pure enzyme to that which may occur in the mitochondria) they may or may not reflect the true concentration.

Attempts were made to detect CPS in lysosomes. When using the postembedding procedure, background interfered markedly. In the preembedding procedure, penetration of the ferritin-anti-CPS conjugate requires breaking to some extent the membranes of cell organelles. Therefore, to avoid background due to CPS diffusion, lysosomes essentially free from mitochondrial contamination (as checked by electron microscopy of samples taken from pellets over their whole depth) were prepared by extreme discriminatory differential centrifugation (where separation rather than yield is the object) (Fig. 8).

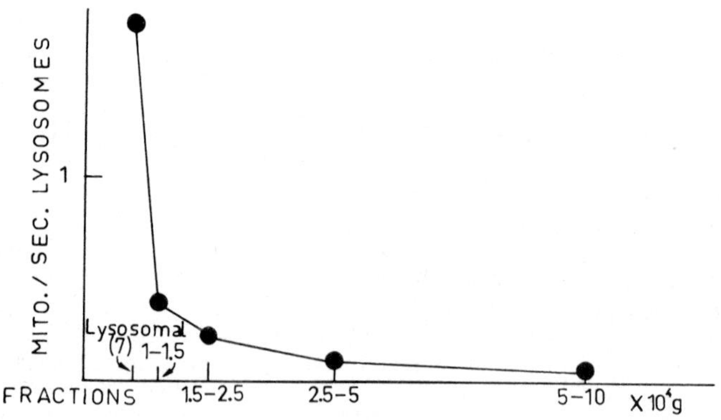


Fig. 8

The 15-25 & 25-50 x $10^3$ g fractions did not show a significant ferritin labeling in lysosomes (Fig. 9). Brain is devoid of CPS; it was added to brain homogenates and incubated for 1 hour. Interestingly, it was apparently taken by some lysosomes while it was not detected in brain mitochondria (Fig. 10).

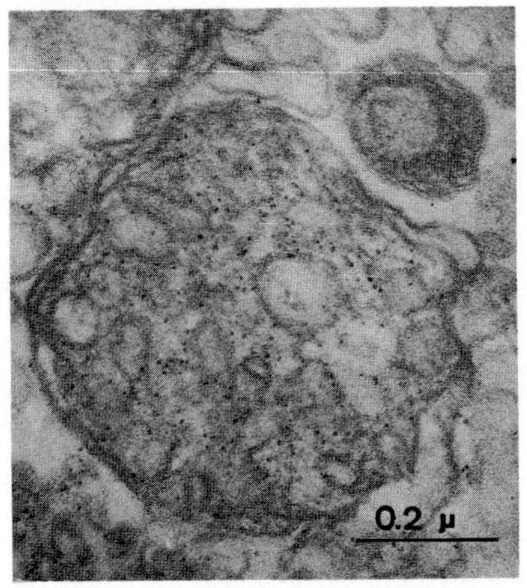


Fig. 9

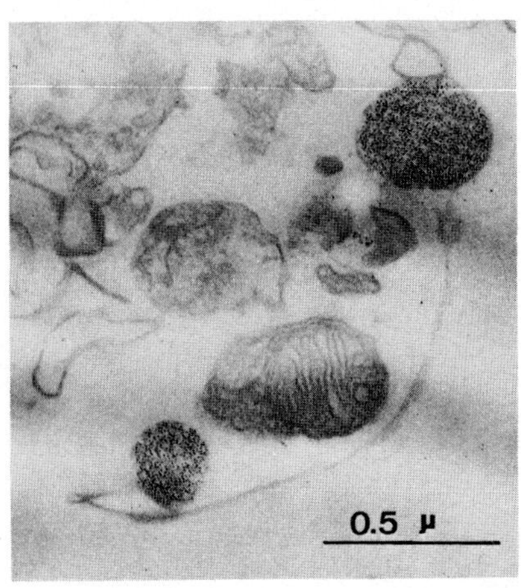


Fig. 10

The possible basis for the discrepancy between the evidence presented above, for the participation of lysosomes in degradation of the bulk of mitochondrial proteins "in vitro", and of proteins of short half-life in cultured cells & the lack of immunologically detectable CPS or fractions thereof in rat liver lysosomes both "in vivo" & "in vitro" together with the apparent uptake of CPS by brain lysosomes are outlined in the scheme.
The uptake of CPS by brain lysosomes may reflect a fast $v_1$, absence of nickases, proteases or catalyzing reactions c, d, e or elastoplastic effects by brain homogenates on a foreign protein. The absence of CPS reactivity "in vivo" or "in vitro" with liver (unless there is proteolysis by mitoplast enzymes), may reflect faster reactions c, d, e over $v_1$ or $v_2$.
The scheme is susceptible of testing by saturating techniques (for reacion $v_1$ and possibly $v_2$) or by lysosomal loading with pure CPS, partially inactivated CPS or with mitoplasts as a source of "nickase" & CPS & particularly by the use of mitochondrial enzymes with shorter $t_{1/2}$ than CPS. At any rate, experiments similar to these already mentioned may go a long way towards solving a number of important problems related to biogenesis as well as degradation of mitochondria, e.g. whether mitochondria possesses equal # of different enzymes, and thus the possi-

bility of clarification of mitochondrial aging & unequivocal quantitation of some enzymes, mechanism(s) of passage of cytosol proteins into mitochondria as well as that of protein degradation thereof and last but most important, explanation & identification of signals which regulate or control the steady state between protein synthesis & degradation.

<u>Scheme for passage of proteases (or proteolytic) initiators) between mitochondria(M) & lysosomes (L)</u>

<u>Immunocytochemical lysosomal labeling</u>

A) Passage (selective) of lysosomal enzymes to mitochondria

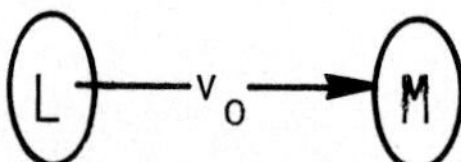


**−**

B) Passage (selective) of mitochondrial proteins to lysosomes.

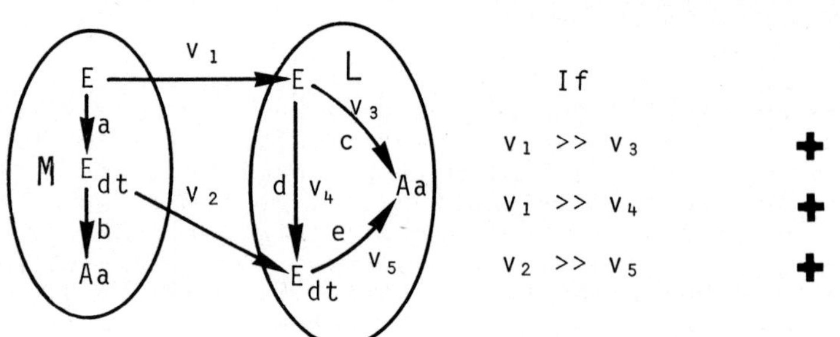


If

$v_1 >> v_3$ **+**

$v_1 >> v_4$ **+**

$v_2 >> v_5$ **+**

where: $v_1$ = velocity of transfer of M enzymes to L
$v_2$ = velocity of transfer of partially denatured M enzymes (either via "nickases" or elastoplastic modification) immunologically reactive, to L
E = native enzymes
$E_{dt}$ = partially denatured enzymes due to "nickase" or elastoplasticity
Aa = aminoacids

## REFERENCES

(1) R.T. Schimke, Control of Enzyme Levels in Mammalian Tissues, <u>Adv. Enzymol</u>. 37, 135 (1973).

(2) A.L. Goldberg & A.C. St. John, Intracellular Protein

Degradation in Mammalian & Bacterial Cells, Ann. Rev. Biochem. 45, 747 (1976).

(3) M. Nicoletti, C. Guerri & S. Grisolía, Turnover of Carbamyl-Phosphate Synthase, of Other Mitochondrial Enzymes & Rat Tissues. Effect of Diet & of Thyroidectomy, Eur. J. Biochem. 75, 583 (1977).

(4) N. Katunuma, E. Kominami, K. Kobayashi, Y. Banno, K. Suzuki, K. Chichibu, Y. Hamagachi & T. Katsunuma. Studies on New Intracellular Proteases in Various Organs of Rat. 1. Purification & Comparison of Their Properties. Eur. J. Biochem. 52, 37 (1973).

(5) O.H. Wieland, On the Mechanism of Irreversible Pyruvate Dehydrogenase Inactivation in Liver Mitochondrial Extract, FEBS Lett. 52, 44 (1975).

(6) V. Rubio & S. Grisolía, Prominent Role of Lysosomes in the Proteolysis of Rat Liver Mitochondria at Neutral pH. FEBS Lett. 75, 281 (1977).

(7) S. Grisolía, J. Rivas, R. Wallace & J. Mendelson, Inhibition of Proteolysis of Cytosol Proteins by Lysosomal Proteases & of Mitochondria of Rat Liver by Antibiotics, Biochem.Biophys.Res.Commun. 77, 367 (1977).

(8) E. Knecht & J. Hernández, Ultrastructural Localization of Polysaccharides in the Vacuolar System of an Established Cell Line. Cell.Tiss.Res. in press (1978).

(9) L. Raijman, Enzyme & Reactant Concentrations & the Regulation of Urea Synthesis, in the Urea Cycle (S. Grisolía, R. Báguena & F. Mayor, eds.) 243-254, Wiley, New York (1976).

(10) M. Marshall & P.P. Cohen, An Immunochemical Study of Carbamyl Phosphate Synthetase, J. Biol. Chem. 236, 718 (1961).

(11) J.G.B. Kwapinski, Methodology of Immunochemical & Immunological Research, Wiley, New York (1972).

(12) H.D. Mayor, J.C. Hampron & B. Rosario, A simple Method for Removing the Resin from Epoxy Embedded Tissue, J. Biophys. Biochem. Cytol. 9, 909 (1961).

(13) J. G. Gamble & A.L. Lehninger, Transport of Ornithine & Citrulline across the Mitochondrial Membrane. J.Biol.Chem. 248, 610 (1973).

(14) C.J. Lusty, Carbamyl Phosphate Synthetase I of Rat Liver Mitochondria. Purification, Properties, & Polypeptide Molecular Weight, Eur. J. Biochem. 85, 373 (1978).

# DEGRADATION OF MITOCHONDRIA IN RETICULOCYTES

T. Schewe and S. Rapoport

Institute of Physiological and Biological Chemistry, Humboldt University, DDR-104 Berlin, GDR

ABSTRACT

The maturation of the reticulocyte to the erythrocyte is characterized by a nearly complete disappearance of mitochondria. Their degradation is preceded by the synthesis of a special lipoxygenase (LOX) unique for the reticulocyte. This enzyme causes lysis of mitochondria by peroxidation of their lipids as well as a multisite inhibition of the respiratory chain by oxidative destruction of iron-sulphur proteins. The LOX has been purified and characterized with respect to its molecular and enzymatic properties. It is a glycoprotein (mol wt 78,000; i.p. 5.6) with functional Fe. The enzyme acts in a suicidal manner by being inactivated during its reaction with substrates. Intensive bleeding anaemia gives rise to the appearance of high amounts of LOX (about 4 mg/ml cells) owing to superinduction.
The degradation of the mitochondria in reticulocytes constitutes the first clear-cut example of the functional role of soluble lipoxygenases in the breakdown of membranes

KEYWORDS

Lipid peroxidation, Lipoxygenase, Reticulocyte maturation, Mitochondrial breakdown, FeS proteins, Respiratory inhibition, Suicidal enzyme reactions.

## INTRODUCTION

The reticulocyte represents an intermediate stage of the red cell which is characterized by the elimination or inactivation of the nucleus on the one hand, and by the presence of functional ribosomes and mitochondria on the other. The maturation, i.e. the transition of the reticulocyte to the erythrocyte is characterized by the disappearance of mitochondria. The breakdown of mitochondria goes hand in hand with a steep decline or complete loss of mitochondrial enzymes (for review see Rapoport and coworkers, 1974). Up till now very little is known about the crucial mechanisms triggering the mass degradation of mitochondria in processes of maturation, involution and ageing in the animal and plant kingdom which are accompanied by a loss of cell respiration, such as the cornification of the skin, maturation of eye lens epithelium, the lignification of wood, the seasonal senescence of leaves etc. The reticulocyte offers the following advantages for the study of the degradation of mitochondria:

1) There is no new-formation of organelles, since an active nucleus and, therefore, RNA synthesis is absent;
2) there is practically no endoplasmic reticulum-except remnants;
3) the degradation of mitochondria during maturation is nearly synchronous;
4) the convenient availability of large amounts of reticulocytes.

From the electronmicroscopic appearance of the mitochondria in intact reticylocytes during the maturation process (Gasko and Danon, 1972; Krause and co-workers, 1972) it is evident, that their degradation proceeds predominately directly in the cytosol rather than in secondary lysosomes.

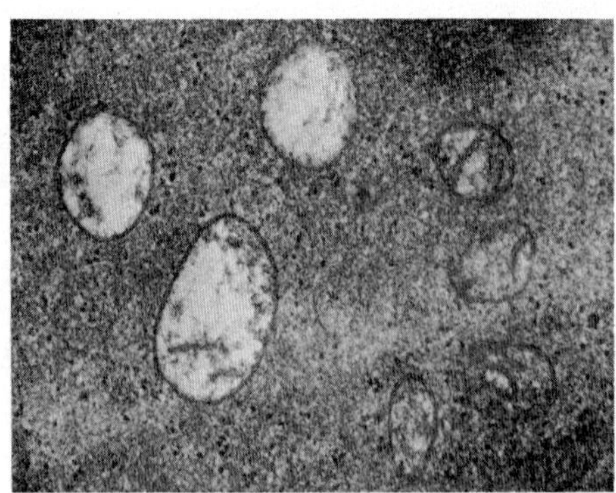

Fig. 1 Mitochondria in a reticulocyte

In Fig. 1 one can see in a reticulocyte partially degraded, swollen and vacuolized mitochondria on the left besides nondegraded, orthodox ones on the right. The share of degradation forms of mitochondria increases with the degree of maturation. In parallel the cell respiration decreases steeply. On account of the rapid and selective disappearance of the cell respiration the occurrence of a factor inactivating the respiratory chain has been predicted and detected (Rapoport and Gerischer-Mothes, 1955). This inhibitor now has been identified as a lipoxygenase (LOX).

## RESULTS AND CONCLUSIONS

### Molecular and Enzymatic Properties of the Reticulocyte LOX

The LOX has been purified from rabbit reticulocytes by ammonium sulphate precipitation, ion exchange chromatography, isoelectric focusing and gel chromatography (Wiesner and co-workers, 1977). Some properties are compiled in Table 1 and Table 2.

TABLE 1 Molecular Properties of the Reticulocyte LOX

- a single polypeptide chain
- mol wt 78,000
- i. p. 5.6
- N-terminal: gly
- C-terminal: ser (gly)
- high percentage of leu and trp
- glycoprotein
- 1.8 moles Fe/mol LOX

TABLE 2 Enzymatic Properties of the Reticulocyte LOX

Fatty acid specificity:

trend of $V_{max}$: $C_{20:3} \geqq C_{18:2} > C_{18:3} \geqq C_{20:4} > C_{18:2}CH_3$

trend of $K_M$: $C_{20:4} < C_{20:3} = C_{18:3} < C_{18:2} < C_{18:2}CH_3$

Linoleic acid hydroperoxide formation:

| | |
|---|---|
| pH optimum: | 7.4 - 7.8 |
| temperature optimum: | 20°C |
| $K_M$: | 0.1 mM |

| Other substrates | Inhibitors |
|---|---|
| all classes of phospholipids<br>mitochondria<br>submitochondrial particles | intermediates of lipid peroxidation (with all substrates); antioxidants (e.g. propyl gallate, α-naphthol); chelators (KCN; Tiron); haemoglobin ($4 \cdot 10^{-4}$ M) |

All polyunsaturated free fatty acids tested are peroxidized. The $K_M$ values tend to decrease with the number of cis-double bonds, whereas the $V_{max}$ values tend to increase with the chain length and the decreasing number of double bonds. Besides free polyene fatty acids, phospholipids, mitochondria and submitochondrial particles are attacked. A striking property of the reticulocyte LOX is the suicidal character of its action: With all substrates listed it is rapidly and irreversibly inactivated in the presence of even traces of oxygen. As an example the inactivation of the reticulocyte LOX during the formation of the hydroperoxide from linoleic acid is shown in Fig. 2. The reaction is accompanied by a loss of activity which is not restored by a subsequent addition of an excess of substrate. By contrast, the activity of the soybean LOX is not affected under identical experimental conditions. Among the inhibitors listed in Table 2 haemoglobin may play a biological role in moderating the LOX attack on mitochondria in reticulocytes (see below).

## Action of the Reticulocyte LOX on Rat Liver Mitochondria

Fig. 3 shows the action of the LOX on isolated rat liver mitochondria. One can recognize disappearance of the outer membranes and the cristae, loss of matrix constituents as well as deformation of the remaining inner membrane. These changes are accompanied by both the release of the matrix enzyme malate dehydrogenase and the formation of malonyl dialdehyde (MDA) which is a secondary product of the peroxidation of membrane phospholipids. The results are presented in detail elsewhere (Halangk and co-workers, 1977, Schewe and co-workers, 1975). The LOX action on mitochondria depends on their functional state. Thus, it is prevented by either ATP or succinate plus ADP, as shown in Table 3. The protective effect is partly counteracted by respiratory inhibitors or uncouplers. One may assume that the protection against LOX is caused by energization of the mitochondrial inner membrane. The basis for the varying susceptibility to

LOX may consist in the state of protein-lipid interactions in the membranes.

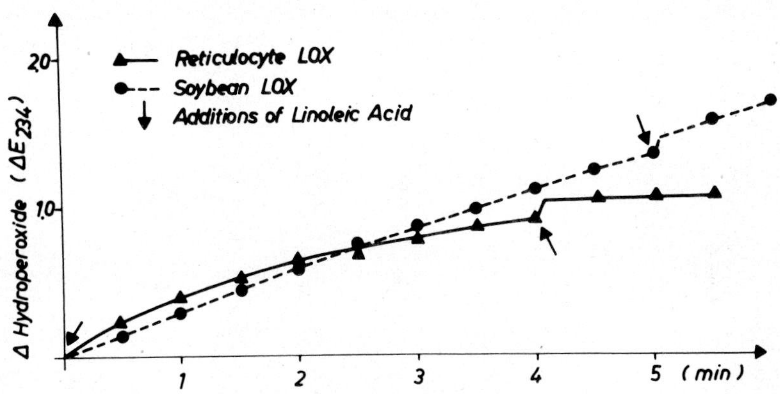


Fig. 2 Inactivation by reaction intermediates of the reticulocyte LOX (T = 29$^{o}$C)

TABLE 3 Dependence of LOX Action on the Functional State of Rat Liver Mitochondria

| Incubation | Method | Action (% MDH release or $\Delta E_{532}$) | Effect (%) |
|---|---|---|---|
| aerobic | MDH release | 90 | |
| aerobic + substrate | " | 2 | - 98 |
| anaerobic | " | 56 | - 38 |
| anaerobic + substrate | " | 10 | - 89 |
| aerobic | MDA formation | 0.182 | |
| aerobic + substrate | " | 0.078 | - 57 |

Incubation 30 min 37$^{o}$C; MDH = malate dehydrogenase MDA = malonyl dialdehyde (determined by the thiobarbituric acid method); substrate: 10 mM succinate plus 1 mM ADP

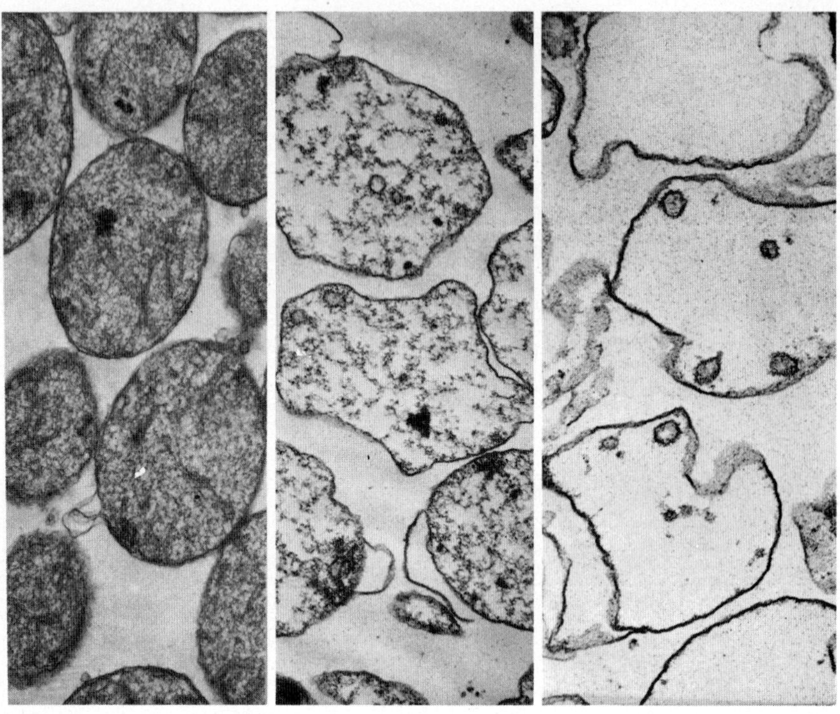

Fig. 3 LOX attack on isolated rat liver mitochondria. From the left to the right: mitochondria after 0, 5 and 15 minutes incubation with LOX at 37°C in 0.3 M buffered sucrose

Erythrocyte ghosts are much more resistant to the LOX than unprotected mitochondria. Unlike mitochondria heat denaturation of the ghosts leads to the enhancement of the formation of MDA by the LOX (Fig. 4) which is also indicative of the role of lipid-protein interaction for the susceptibility to LOX. The resistance of the red cell membrane may be the basis for the selective action of the LOX on the mitochondrial membranes during maturation. Erythrocyte cytosol exerts a moderating effect on the LOX action in a model system containing isolated rat liver mitochondria (Fig. 5). This may explain the fact that maturational changes in vivo are slower than those in vitro.

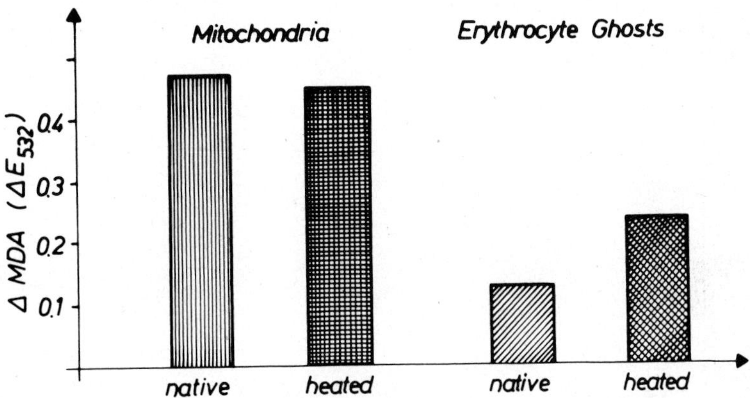


Fig. 4 MDA formation by reticulocyte LOX from rat liver mitochondria and erythrocyte ghosts. Identical protein content of the samples and identical incubation conditions.

## Action of the LOX on Electron Transfer Particles (ETP)

With ETP the reticulocyte LOX causes an irreversible inhibition of the respiratory chain at at least three different sites (Fig. 6). They are the two iron-sulphur regions in the complexes I and II (designated as RF action) as well as the cytochrome oxidase (designated as RC action). The actions of the reticulocyte LOX on ETP and other respiratory enzymes have been presented in detail elsewhere (Schewe, Hiebsch and Rapoport, 1972; Wiesner and Rapoport, 1973). As we could now find out, the LOX causes in ETP a loss of acid-labile sulphur which parallels the respiratory inhibition (Table 4). Therefore, the respiratory inhibition in the complexes I and II is probably due to the destruction of iron-sulphur centres, presumably by intermediate radicals of the reaction of the

LOX with membrane phospholipids in the presence of oxygen.

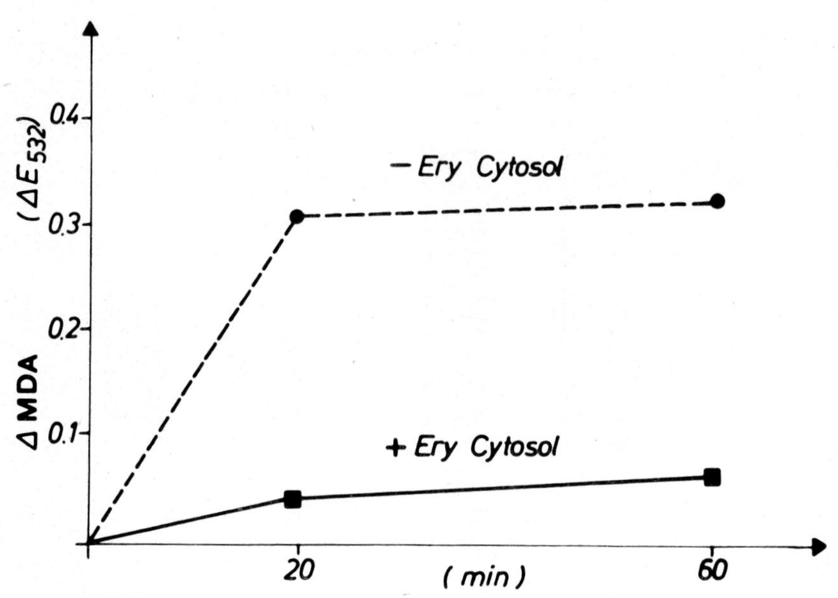


Fig. 5 MDA formation from rat liver mitochondria (simulated cellular conditions of the reticulocyte)

TABLE 4 LOX Effect on FeS Proteins of ETP

| | | Acid-labile S (nmoles/mg) | | | Inhibition (%) | |
|---|---|---|---|---|---|---|
| Exp. | LOX (µg/mg ETP) | ETP | ETP + LOX | Δ | NADH oxidase | Succ. - cyt. c red. |
| I | 8.1 | 10.3 | 7.8 | 2.5 | 92 | — |
| II | 7.5 | 10.4 | 6.9 | 3.5 | 97 | 54 |
| | 3.2 | 11.9 | 10.1 | 1.8 | 61 | 25 |

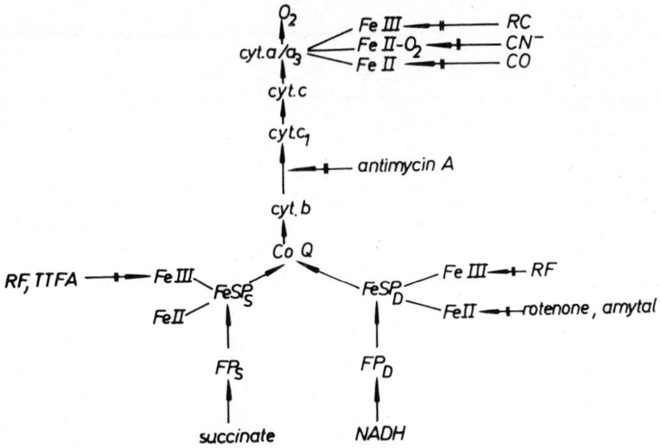

Fig. 6 The action sites RF and RC of the LOX in the respiratory chain

The inhibition of the cytochrome oxidase which shows a different kinetics of its appearance has to be attributed to another mechanism of action possibly via a damage to the haeme system. The amount of LOX necessary for the respiratory inhibition is very low. 1 mg of LOX is sufficient to inhibit about 500 mg of ETP protein. On the molar basis the LOX is more efficient than rotenone, one of the most powerful inhibitors. The reticulocyte LOX is more efficient by about one order of magnitude with respect to the respiratory inhibition as compared with soybean LOX. In former publications (e.g. Schewe and co-workers, 1977) we had claimed that the LOX lysing mitochondria (Schewe and co-workers, 1975) and the respiratory inhibitors were different factors. This assumption was based mainly on observations of distinct effects of some inhibitors and manipulations on the various activities. However, now we have obtained a variety of proofs for the chemical and functional identity such as identity of mol wt and i.p., identical behaviour in electrophoresis, ion exchange, affinity chromatography and immunoprecipitation. Furthermore, the activities behave identically with respect to their dependence on oxygen, their aerobic inactivation by linoleic acid or phospholipids and the binding to submitochondrial particles.

## Synthesis and Biological Dynamics of the Reticulocyte LOX

The LOX is absent in normal blood as judged both by enzyme assay and immunologically but appears during bleeding anaemia, as seen from Fig. 7. The LOX appears generally on the fourth day and increases steeply up to the sixth day approximately in parallel to the percentage of reticulocytes. Further bleeding further increases the LOX the content of which reaches values of 4 mg/ml cells. After cessation of the bleeding the LOX persists for a long time. The appearance of large amounts of LOX appears to represent a superinduction.

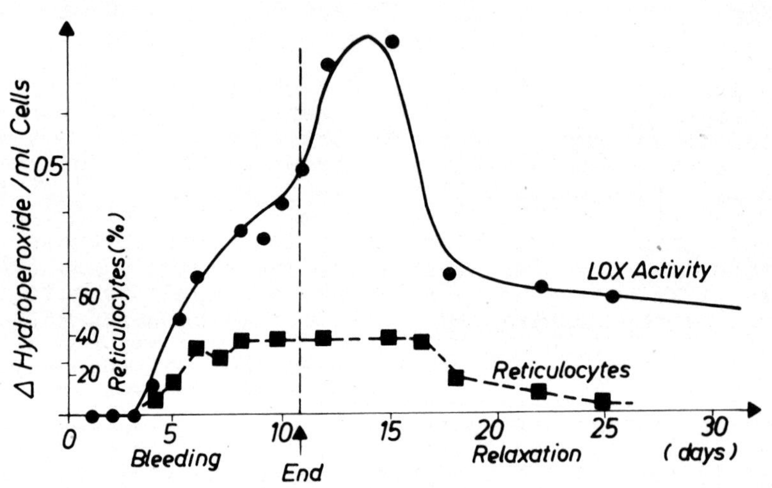


Fig. 7 LOX activity in bleeding anaemia

Fig. 8 shows schematically the results of studies on cell populations separated by buoyant density centrifugation according to their age. Some maturational criteria are plotted versus the decreasing RNA content which is a measure of the maturational stage of the reticulocyte. In very young, ribosome-rich cells as well as in bone marrow cells no LOX is detectable.

The maximal activity-here expressed by the inhibitory activity-is reached at the very time of the steepest decline of mitochondrial and other maturation-dependent parameters. Subsequently, the LOX disappears again. This may be accounted for the suicidal nature of the enzyme reaction. From this behaviour it is evident that the LOX appears at a rather late stage of maturation.

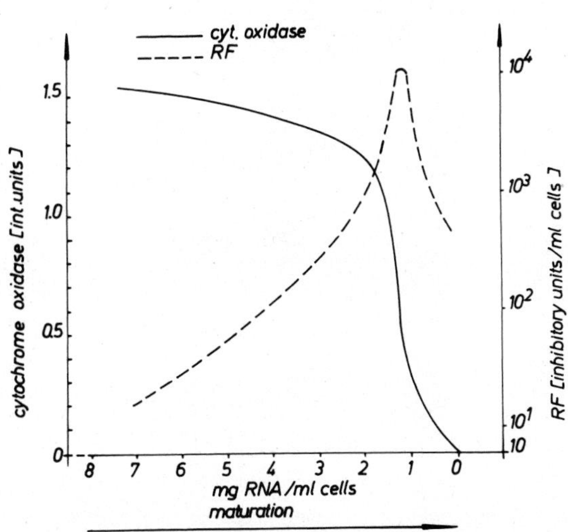


Fig. 8 Biological dynamics of cytochrome oxidase (full line) and LOX (measured by the inhibition of the NADH oxidase system of ETP-broken line) during the maturation of the reticulocyte. The curve of cytochrome oxidase is also representative for other maturation-dependent enzymes (δ-ALA synthetase, GOT, glutaminase, MDH, DNase, PPase, glucokinase). The age of the cell population increases with decreasing RNA content.

The synthesis of LOX takes place in the reticulocyte as demonstrated by the incorporation of labelled leucin (Fig. 9). From this result it follows that there exists in reticulocytes a stable mRNA or a precursor of it. In any case, the synthesis of the LOX may be the key event in triggering the degradation of mitochondria and, hence, the maturation.

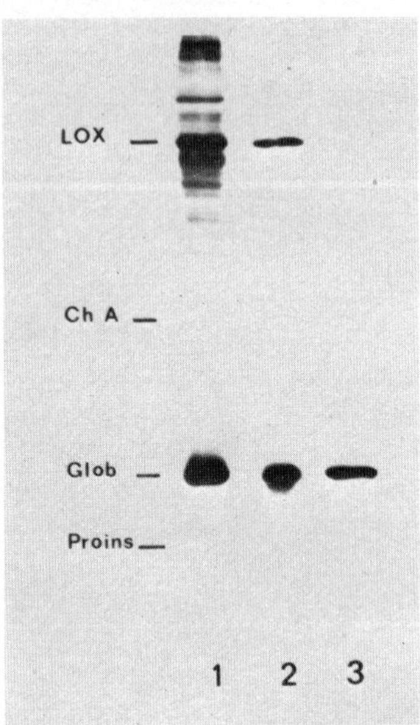


Fig. 9 Autoradiogramme of the polypeptides of reticulocyte cytosol separated by SDS electrophoresis after incorporation of $^{3}$H-labelled leucin in intact reticulocytes. 1 = crude protein fraction, 2 = precipitation by anti-LOX serum, 3 = control precipitation by normal serum (unspecific binding of haemoglobin)

## The Role of the LOX in the Degradation of Mitochondria in Reticulocytes.

The consequences of the LOX action on mitochondria are compiled in Table 5.

In this manner, the degradation process-once initiated by the LOX- is apt to have a self-accelerating character leading to a catastrophe cascade with a complete breakdown of mitochondria.

The degradation of the mitochondria of reticulocytes constitutes the first clear-cut example of the functional role of soluble lipoxygenases in the breakdown of membranes. A similar role seems to play the lipid peroxidation in the breakdown of the membranes of the endoplasmic reticulum of liver after deprivation of a substrate to be hydroxylized (Kagan and co-workers, 1974). Even in the degradation of chloroplasts during seasonal withering and abscission analogous events should be anticipated.

TABLE 5 LOX Actions on Mitochondria and Their Consequences

| Action | Consequences |
|---|---|
| peroxidation of mitochondrial membrane lipids | 1) damage to membrane-linked functions<br>2) mediation of penetration of itself and of other degrading factors from the cytosol |
| respiratory inhibition | 1) interruption of ATP synthesis<br>2) greater susceptibility to proteinases and phospholipases (Luzikov and Romashina, 1972)<br>3) accumulation of free fatty acids with further damage to mitochondrial functions |

ACKNOWLEDGEMENTS

The authors are grateful to W. Halangk, Ch. Hiebsch, M. Höhne, D. Klatt, P. Ludwig, Ch. Tannert, B. Thiele, and R. Wiesner for the participation in the studies presented here. Furthermore, the authors are indebted to W. Krause, Humboldt-University, Pathological Institute, Division of Electronmicroscopy, Berlin, GDR for the kind performance of the electronmicrographs.

REFERENCES

Gasko, O., and D. Danon (1972). Deterioration and disappearance of mitochondria during reticulocyte maturation. Expl. Cell Res., 75, 159-169.

Halangk, W., T. Schewe, Ch. Hiebsch, and S. Rapoport (1977). Some properties of the lipoxygenase from rabbit reticulocytes. Acta Biol. Med. Germ., 36, 405-410.

Kagan, V. E., C. V. Kotelevtsev, and Yu. P. Kozlov (1974). The role of enzymatic peroxidation of phospholipids in the mechanism of breakdown of the membranes of endoplasmic reticulum. Dokl. Akad. Nauk SSSR (Proc. Acad. Sci. USSR), 217, 213-216 (in Russian).

Krause, W., H. David, I. Uerlings, and S. Rosenthal (1972). Veränderungen der Mitochondrien-Ultrastruktur von Kaninchenretikulozyten im Reifungsprozeß. Acta Biol. Med. Germ., 28, 779-786.

Luzikov, V. N., and L. V. Romashina (1972). Studies on stabilization of the oxidative phosphorylation system. II. Electron transfer-dependent resistance of succinate oxidase and NADH oxidase of systems of submitochondrial particles to proteinases and cobra venom phospholipase. Biochim. Biophys. Acta, 267, 37.

Rapoport, S., and W. Gerischer-Mothes (1955). Biochemische Vorgänge bei der Erythrozytenreifung: Über einen Hemmstoff des Succinatoxydase-Systems in Retikulozyten. Hoppe-Seyler's Z. Physiol. Chem., 302, 167-178.

Rapoport, S. M., S. Rosenthal, T. Schewe, M. Schultze, and M. MÜLLER (1974). The metabolism of the reticulocyte. In H. Yoshikawa and S. M. Rapoport (Ed.), Cellular and Molecular Biology of Erythrocytes, University of Tokyo Press, Tokyo. pp. 93-141.

Schewe, T., Ch. Hiebsch, and S. Rapoport (1972). Biochemische Vorgänge bei der Erythrozytenreifung. Weitere Ergebnisse zum Wirkort des Atmungshemmstoffes F aus Retikulozyten in der Atmungskette. Acta Biol. Med. Germ., 29, 189-206.

Schewe, T., W. Halangk, Ch. Hiebsch, and S. M. Rapoport (1975). A lipoxygenase in rabbit reticulocytes which attacks phospholipids and intact mitochondria. FEBS Letters, 60, 149-152.

Wiesner, R., and S. Rapoport (1973). Wirkung von Hemmstoff RC aus Kaninchenretikulozyten auf lösliche und partikuläre Zytochromoxydase der Rinderherzmitochondrien. Acta Biol. Med. Germ., 31, 289-304.

Wiesner, R., Ch. Tannert, G. Hausdorf, T. Schewe, and S. Rapoport (1977). Reinigung und Charakterisierung des Atmungshemmstoffes RF aus Kaninchenretikulozyten. Acta Biol. Med. Germ., 36, 393-403.

# SYNTHESIS AND TURNOVER OF THE HEPATIC PLASMA MEMBRANE

W. Howard Evans and Joan A. Higgins
National Institute for Medical Research, Mill Hill, London NW7 1AA, U.K. and Section of Cytology, Yale University, Connecticut, U.S.A.

## ABSTRACT

The synthesis and turnover of the plasma membrane of cells of secretory tissues is a function of cellular metabolic activity. In the hepatocyte, secretory and uptake processes occur predominantly at the blood-sinusoidal plasma membrane region. However, most liver plasma membrane fractions used in turnover studies derive mainly from the lateral and bile canalicular region and thus cannot reflect accurately the overall extent of synthesis and turnover occurring at this organelle. Proteins and glycoproteins may be inserted into one or more of the major functional domains, and the lateral relocation of plasma membrane components between domains is affected by junctions especially tight junctions. To gain further insight into the mechanism of plasma membrane assembly, the transverse organisation of the phospholipids across the bilayer of blood sinusoidal, contiguous and bile canalicular plasma membranes and microsomal membranes was determined; a complex biogenetic relationship exists between these membrane systems.

## INTRODUCTION

The plasma membrane is one of the most complex and multifunctional of the cell membrane structures. This is especially the case in cells that make up tissues and organs, because plasma membrane structure and function varies regionally according to intracellular organisation and the nature of the external milieu. Thus, in cells exhibiting functional polarity the composition and transverse topographical arrangement of proteins and lipids of plasma membranes may vary according to their location at the various functional domains, a feature not so obvious in cells which normally occur singly, and on which most studies of plasma membranes have been performed. In general, the mammalian plasma membrane lacks major intrinsic biogenetic capability, having to derive its proteins and lipids from intracellular sites of synthesis. Studies on the synthesis and turnover of hepatocyte plasma membranes must thus take into account the functional and molecular heterogeneity of the three main plasma membrane regions, and, especially, the biogenetic relationship between sites of synthesis of membrane components and the various possible points of insertion into the plasma membrane. Further, since rates of synthesis and turnover are calculated using membrane fractions prepared from tissues labelled with appropriate isotopic precursors, the purity, representiveness and regional derivation of these membranes is of prime importance. The present account is a critique of the methods used to evaluate the synthesis and turnover of hepatic plasma membranes and focusses on the need to consider the results in the context of the three major functional domains. Also, a more direct approach is described for studying the transverse orientation of the phospholipids of the hepatocyte's plasma membrane functional domains and microsomal membranes, so providing further insight into the biogenetic relationships between these two membrane components.

## The Hepatocyte Plasma Membrane : a Dynamic Functional Mosaic

The hepatocyte is extremely complex metabolically being active both in secretion into the blood system, and in the uptake of metabolites brought to the liver by the portal venous system. Recently, the liver has been shown to be the main site for the degradation of a range of glycoproteins containing a terminal galactose residue (1, 2). Thus, the blood sinusoidal plasma membrane region of the hepatocyte participates in events that result in membrane insertion occurring during secretion and membrane depletion during uptake of components circulating in the blood. The extent of involvement of individual hepatocytes appears to vary according to their position in the liver lobule and hence their proximity to arterial and venous (portal) systems (3). Linking this dynamic interplay between secretory and uptake mechanisms is the possibility that there is direct recycling of internalised membranes back to the plasma membrane, or degradation following fusion with lysosomes (Fig. 1), processes that appear to occur in macrophages (4), fibroblasts (5) and hepatoma tissue culture cells (6). A further possibility occurring in secretory tissues is the interaction of endocytosed macromolecules with Golgi cisternae, an intracellular pathway that is influenced by the net charge of the macromolecule (7). Thus, results of studies on the synthesis and degradation of the blood sinusoidal plasma membrane region of the hepatocyte are highly influenced by the cell's secretory activity, and the nature of the components in the liver blood supply.

Two further functional domains of the hepatocyte may be distinguished: the bile cannicular region participating in the release into the cannicular spaces of bile and other metabolites (previously taken up at the sinusoidal plasma membrane) and a flattened membrane region contiguous with adjacent hepatocytes and characterised by the presence of intercellular junctions (Table 1). Since the plasma membrane is physically continuous and there may be lateral diffusion of protein, glycoprotein and lipid components within the plane of the membrane, then events occurring at each of these functional domains, especially the high metabolic activity at the blood-sinusoidal region, may have inductive short and long range effects on the other regions.

TABLE 1 Properties of Liver Plasma Membrane Fractions

| Surface region | Morphological and physical properties | Biochemical properties |
|---|---|---|
| Blood-sinusoidal | Microvillar; accounts for 40–50% of surface area. Plasma membranes are vesicles of density 1.11–1.14 in sucrose and are recovered mainly from microsomal fractions | Identified by perfusing liver with $^{125}$I-ligands. High sialic acid and glycoprotein content. Intermediate specific activities of piasma-membrane marker ecto-enzymes, but adenylate cyclase highly-activated by glucagon. Difficult to separate from Golgi membranes |
| Contiguous (lateral) | Flat; accounts for 30–40% of surface area. Plasma membranes are of density 1.16–1.18. Intercellular junctions present | Lowest specific activities of plasma-membrane marker ecto-enzymes. Low glycoprotein and glycolipid content. Gap-junctional polypeptides prominent |
| Canalicular | Microvillar; accounts for ~10% of surface area. Plasma membranes isolated as canalicular complexes (density 1.16–1.18) or vesicles (density 1.13) | Highest specific activities of plasma membrane marker ecto-enzymes, cholesterol and sphingomyelin. Little activation of adenylate cyclase by glucagon. High sialic acid and glycoprotein content |

## Synthesis and Turnover of Hepatic Plasma Membrane Fractions

The dynamic state of hepatic membrane proteins and lipids has become evident from studies on microsomal fractions. These have shown that specific proteins are synthesised and degraded at widely differing rates. Such information on the synthesis and degradation of plasma membrane enzymes is sparse e.g. NAD glycohydrolase (8), and major efforts are confined to the determination of the rates of turnover of overall membrane fractions. When precautions are taken to minimise isotope reutilisation, liver plasma membrane proteins have an average half life of 41-43 h (using sodium $^{14}$C-carbonate (9) or guanidino-$^{14}$C-arginine (10). A more rapid turnover of large proteins and of glycoproteins appears to occur (10, 14). Using $^{14}$C-glucosamine or $^{3}$H-fucose, half-lives of 37 h and 33 h respectively for plasma membrane glycoproteins were obtained (11, 12). Other studies have identified a group of plasma membrane glycoproteins, especially fucoproteins, that appear to turnover much more rapidly (13). In regenerating liver, and in hepatoma, the half-lives of glycoproteins (9, 12) and protein (6) are significantly longer (8), presumably reflecting a lower degree of secretory activity owing to cell growth and proliferation.

Such investigations are subject to the criticism that the plasma membranes used originated mainly from the contiguous and bile canalicular regions, since they were prepared from 'low speed' pellets of tissue homogenates. Thus, the highest metabolic activities that are associated with the blood-sinusoidal plasma membranes and are reflected in higher turnover and possibly re-utilisation of isotopic markers are only partly accounted for depending on the persistence of membrane fragments from this region in the collected fractions. Blood-sinusoidal plasma membranes are recovered mainly in the microsomal pellet (15, 16), which is examined only rarely in liver plasma membrane studies. Furthermore, when comparing rates of turnover of plasma membrane components of liver with hepatoma or cultured liver cells, the differentiation of the surface of 'normal' hepatocytes into physiologically distinct regions highlights the probability that the plasma membrane of 'dedifferentiated' cells will show deviant fractionation behaviour.

Little is known of the mechanism of plasma membrane degradation, the enzymes involved, and their locus of action.

## Membrane Biogenesis in Multifunctional Cells

In a polar multifunctional cell, the possibility arises that membrane proteins and lipids may be inserted into the plasma membrane at one or more regions. In the hepatocyte, two extreme possibilities (17) may be involved. In the first, components destined to become structural parts of the plasma membrane are inserted directly into the position where they are functional and are restricted from moving laterally into the other functional domains. Lateral movements may be restricted by the association, for example, of transmembrane proteins with cytoskeletal elements, or receptor proteins with ligands. Gap junctional proteins are confined to the lateral plasma membrane region, owing to their tight association with corresponding proteins on contiguous cells, and such a lateral containment model is applicable to other intercellular junctional components (desmosomes, tight junctions). A second alternative postulates insertion into a single domain followed by redistribution by lateral movement in the plane of the membrane. Since some membrane proteins can move laterally, whereas others do not (16), this may explain the different concentrations of, for example, glycoproteins on the apical and lateral sides of epithelial cells (19). However, in considering the plausibility of lateral redistribution, the presence in plasma membranes of tight junctions as restrictive barriers must be

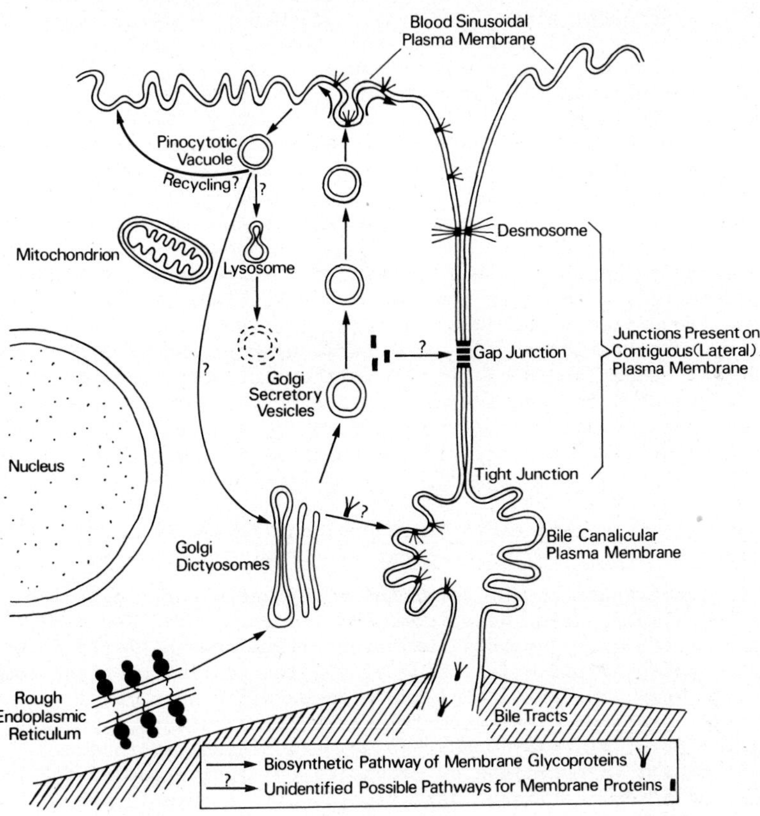


Fig. 1. Pathways of insertion and removal of components of the three major functional domains of the hepatic plasma membrane.

Proteins, synthesised at the rough endoplasmic reticulum, are glycosylated in the Golgi apparatus and inserted into the blood sinusoidal plasma membrane. Interiorised membrane may recycle to the plasma membrane, fuse with lysosomes or Golgi dictyosomes. The tight junctions may form a barrier restricting lateral diffusion into the bile canalicular membrane.

considered. Tight junctions are intercellular permeability barriers completely surrounding cells and composed of fibrils of closely packed intramembranous particles sealing off the extracellular space (20). In hepatocytes, insertion of glycoproteins into the plasma membrane during interaction of Golgi membranes with the sinusoidal plasma membrane, followed by lateral diffusion to contiguous regions may explain the biochemical findings of higher glycoprotein levels and ectoenzymes at the sinusoidal plasma membrane than in contiguous plasma membranes (14). However, the highest amounts of glycoproteins are found at the bile canalicular plasma membrane, a region far removed geographically from a route of insertion involving Golgi-sinusoidal plasma membrane interaction and furthermore, a region enclosed by the tight junction. Although the biogenesis of plasma membrane glycoproteins such as 5'-nucleotidase or nucleotide pyrophosphatase, can be interpreted reasonably in the context of a route from the endoplasmic reticulum to Golgi apparatus to sinusoidal plasma membrane to contiguous plasma membrane (see Fig. 1), their presence at highest concentration at the bile canalicular plasma membrane cannot yet be explained easily. A further influence on the rates of turnover of bile canalicular membrane is the loss of membrane components especially ectoenzymes into bile, a process for which the detergent-like property of bile salts is a contributory factor (21, 26).

## Phospholipid arrangements in Microsomal and Plasma Membrane Fractions - Biogenetic Implications

A more direct approach for determining changes occurring in membrane topography during secretion, and the structural and biogenetic relationship between the three plasma domains is to determine the phospholipid orientation of the interacting membrane components. Membrane vesicles of known orientation in a rat liver microsomal fraction (22) (ascertained by ribosome topography and localisation of glucose-6-phosphatase) and plasma membrane fractions (23) (ascertained by immunological inhibition of the ectoenzyme 5'-nucleotidase) were treated with phospholipase C from Clostridium welchii under conditions in which vesicle content and permeability were monitored. The composition of the phospholipids of the outer leaflet of the membrane bilayers was determined by measuring hydrolysis under conditions in which the vesicles remained intact. In control experiments, membrane lipids were extracted and dispersed or membrane vesicles were opened experimentally in order to demonstrate that the results with closed vesicles were not a consequence of the specificity of the phospholipase C. The orientation of the phospholipids in microsomal and functionally-characterised plasma membrane fractions from these investigations is shown in Fig. 2. The results indicate:

a) a complex biogenetic relationship exists between intracellular membranes and the plasma membranes. Clearly, the transverse

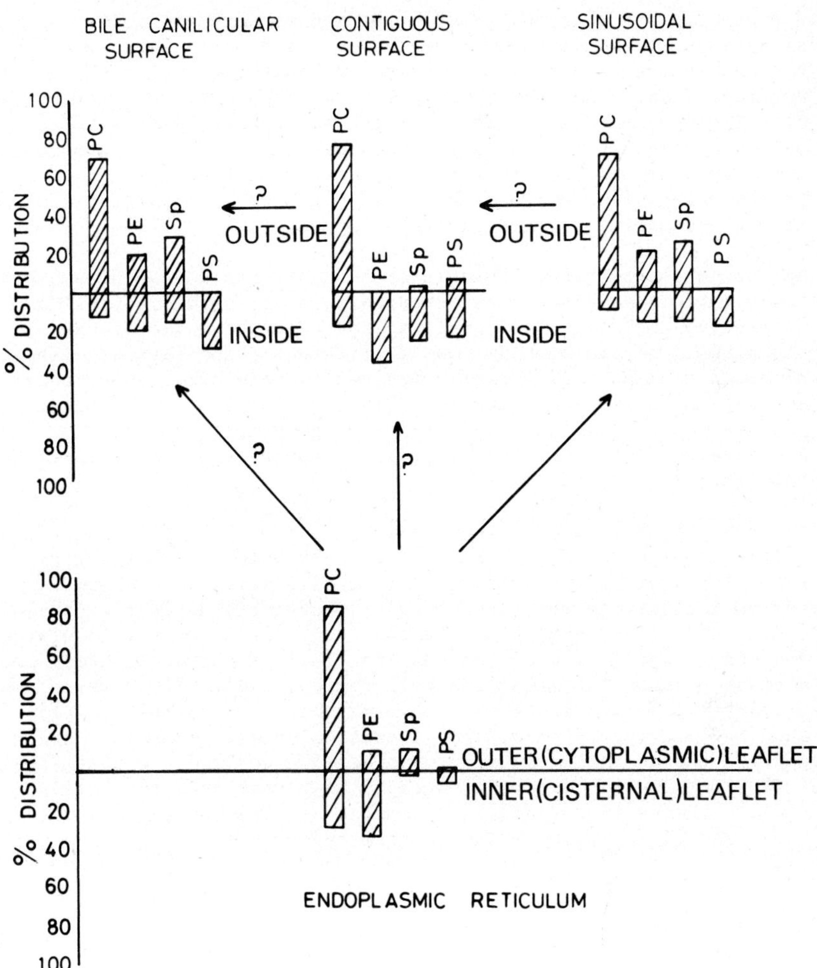


Fig. 2. Transverse distribution of phospholipids of liver plasma membrane and microsomal fractions as determined by phospholipase C treatment.

The arrows indicate possible biogenetic pathways from the site of synthesis in the smooth endoplasmic reticulum directly or indirectly by lateral movement to the various plasma membrane functional domains.

---

distribution of phospholipids in the endoplasmic reticulum and the sinusoidal plasma membrane revealed by the present experiments are not consistent with a simple fusion mechanism, in which the inner surface of the endoplasmic reticulum would become the outer surface of the plasma membrane. Thus, if the endoplasmic reticulum contributes directly to the plasma membrane, changes in phospholipid topography as well as membrane composition must occur. Knowledge of the orientation of phospholipids in the Golgi apparatus would help to determine aspects of such changes.

b) although the blood sinusoidal and bile canalicular plasma membrane regions have a similar phospholipid arrangement (possibly reflecting their microvillar form) the contiguous (junction-containing) plasma membrane has a different organisation. Thus, although the three plasma membrane regions are continuous, there must be mechanisms that ensure different composition and molecular arrangements. Such differences are, for example, further accentuated at the specialised plasma membrane regions comprising the gap junctions (24).

c) the phospholipids in the outer bilayer of the bile canalicular plasma membrane do not correspond closely with the phospholipid composition of bile (Table 2). Since bile salts were shown to remove selectively the outer bilayer phospholipids from erythrocytes (25) it has been proposed that the same mechanism involving selective dissolution of phosphatidyl choline and ectoenzymes may account in part for mammalian bile composition (26). However, the present results indicate that this may be an oversimplification, and that due regard should be given to a direct metabolic involvement in the formation of bile phosphatidyl choline and cholesterol (27). Information on the topography of plasma membrane cholesterol in the various functional domains of the hepatocyte should prove useful in further testing this hypothesis.

TABLE 2 Phospholipid Composition of Hepatic Canalicular Plasma Membrane and the Outer Phospholipid Leaflet in Relation to Bile

| Phospholipid | Bile canalicular plasma membrane (27) | Outer leaflet of * bile canalicular membrane(23) | Bile (21) |
|---|---|---|---|
| PC | 32.5 | 57.0 | 81.5 |
| PE | 14.0 | 17.0 | 7.4 |
| SP | 24.4 | 23.0 | 3.9 |
| PS | 7.8 | 0 | 2.5 |
| PI | 7.0 | 0 | 3.4 |
| Others | 14.3 | - | 1.3 |

* Phospholipids hydrolysed by phospholipase C under conditions in which vesicles remain intact.

## REFERENCES


1. G. Ashwell & A.G. Morell, Adv. Enzymol. 41, 99 (1974).
2. H. Tolleshaug, T. Berg, M. Nilsson & K.R. Norum, Biochim. Biophys. Acta 499, 73 (1977).
3. A.B. Novikoff, J. Histochem. Cytochem. 7, 240 (1959).
4. R.M. Steinman, S.E. Brodie & Z.A. Cohn, J. Cell Biol. 68, 665 (1976).
5. P. Tulkens, Y.-J. Schneider & A. Trouet, Biochem. Soc. Trans. 5, 1809 (1977).
6. D. Doyle, H. Baumann, B. England, E. Hou & J. Tweto, J. Biol. Chem. 253, 965 (1978).
7. M.G. Farquhar, J. Cell Biol. 77, R35 (1978.
8. K.W. Bock, P. Siekevitz & G.E. Palade, J. Biol. Chem. 246, 188 (1971).
9. R. Tauber & W. Reutter, Europ. J. Biochem. 83, 37 (1978).
10. I.M. Arias, D. Doyle & R.T. Schimke, J. Biol. Chem. 244, 3303 (1969).

11. T. Kawasaki & I. Yamashina, Biochem. Biophys. Acta 225, 234 (1971).

12. A.E. Sirica, P.J. Goldblatt & J.F. McKelvy, J. Biol. Chem. 252, 5895 (1977).

13. P. Vischer, Dissertation Universitat Freiburg (1977).

14. J.W. Gurd & W.H. Evans, Europ. J. Biochem. 36, 273 (1973).

15. M.H. Wisher & W.H. Evans, Biochem. J. 146, 375 (1975).

16. F. Carey & W.H. Evans, Biochem. Soc. Trans. 5, 103 (1977).

17. W.H. Evans, Biochem. Soc. Trans. 4, 1007 (1976).

18. D.F.H. Wallach, in Structural and Kinetic Approach to Plasma Membrane Functions (ed. by C. Nicolau & A. Paraf), Springer-Verlag, Heidelberg (1977).

19. G. Bennett, C.P. Leblond & A. Haddad, J. Cell. Biol. 60, 258 (1974).

20. L.A. Staehlin, J. Cell. Science 13, 763 (1973).

21. W.H. Evans, T. Kremmer & J.G. Culvenor, Biochem. J. 154, 589 (1976).

22. J.A. Higgins & R.M.C. Dawson, Biochim. Biophys. Acta 470, 342 (1977).

23. J.A. Higgins & W.H. Evans, Biochem. J. 173, in press (1978).

24. W.H. Evans & J.W. Gurd, Biochem. J. 128, 691 (1972).

25. D. Billington, R. Coleman & Y.A. Lusak, Biochem. Biophys. Acta 466, 526 (1977).

26. R. Coleman, G. Holdsworth & O.S. Uyvoda, in Membrane Alterations as Basis of Liver Injury (ed. by H. Popper, L. Bianchi & W. Reutter) M.T.P. Press, Lancaster, U.K. (1977).

27. T. Kremmer, M.H. Wisher & W.H. Evans, Biochim. Biophys. Acta 455, 655 (1976).

28. D.H. Gregory, Z.R. Vlahcevic, P. Schatzki & L. Swell, J. Clin. Invest. 55, 105 (1975).

# AUTHOR INDEX

# SUBJECT INDEX

The page numbers refer to the first page of the contribution in which the index term appears.